Stem Cells and Cloning
Second Edition

Kelly A. Hogan

University of North Carolina

Michael A. Palladino, Series Editor

Monmouth University

San Francisco   Boston   New York
Cape Town   Hong Kong   London   Madrid   Mexico City
Montreal   Munich   Paris   Singapore   Sydney   Tokyo   Toronto

Editor-in-Chief: Beth Wilbur
Executive Director of Development: Deborah Gale
Acquisitions Editor: Becky Ruden
Project Editor: Nina Sparer
Managing Editor: Michael Early
Production Supervisor: Lori Newman
Production Management: Katie Ostler and Margot Schmidt, Black Dot Group
Copy Editor: Black Dot Group
Compositor: Black Dot Group
Design Manager: Mark Ong
Interior Designer: Black Dot Group
Cover Designer: Seventeenth Street Studios
Illustrators: Black Dot Group
Photo Researcher: David Chavez
Director, Image Resource Center: Melinda Patelli
Image Rights and Permissions Manager: Zina Arabia
Image Permissions Coordinator: Elaine Soares
Manufacturing Buyer: Michael Penne
Executive Marketing Manager: Lauren Harp
Text and Cover Printer: Edwards Brothers Malloy

Cover Image: Colored scanning electron micrograph (SEM) of a human embryo at the 10-cell stage on the tip of a pin. The ball of cells (yellow) of the embryo is known as the morula, a cluster of almost identical, rounded cells; each contains a central nucleus. This 10-cell embryo is about three days old. Dr. Yorgas Nikas/Photo Researchers.

Copyright ©2009 Pearson Education, Inc., publishing as Pearson Benjamin Cummings, 1301 Sansome St., San Francisco, CA 94111. All rights reserved. Manufactured in the United States of America. This publication is protected by Copyright and permission should be obtained from the publisher prior to any prohibited reproduction, storage in a retrieval system, or transmission in any form or by any means, electronic, mechanical, photocopying, recording, or likewise. To obtain permission(s) to use material from this work, please submit a written request to Pearson Education, Inc., Permissions Department, 1900 E. Lake Ave., Glenview, IL 60025. For information regarding permissions, call (847) 486-2635.

Many of the designations used by manufacturers and sellers to distinguish their products are claimed as trademarks. Where those designations appear in this book, and the publisher was aware of a trademark claim, the designations have been printed in initial caps or all caps.

Pearson/Benjamin Cummings is a trademark, in the U.S. and/or other countries, of Pearson Education, Inc. or its affiliates.

ISBN 0321590023 / 9780321590022

4 5 6 7 8 9 10—V0SV—16 15 14
www.pearsonhighered.com

Contents

Introduction — 1

Stem Cell Basics — 2
    What Is a Stem Cell? 2
    Where Do Stem Cells Come from? 4

Embryonic Stem Cells (ESCs) — 7
    Characteristics of Embryonic Stem Cells 7
    Tests for Pluripotency, the Ability to Form Any Body Tissue 9
    What Are the Possible Uses for Embryonic Stem Cells? 11
    What Are the Problems with Embryonic Stem Cells? 13

Adult-Derived Stem Cells (ASCs) — 14
    Characteristics of Adult-Derived Stem Cells 14
    Adult-Derived Stem Cells Are Already in Use in Regenerative Medicine 16

A Multipronged Approach to Curing Diseases — 19

Cloning — 22
    What Is Cloning? 22
    Purposes of Cloning—Why Do We Want to Clone? 24
    Reproductive Cloning 24
    Therapeutic Cloning 28

Bioethics of Stem Cells and Cloning 29
    Bioethics of Stem Cell Research   29

    Alternatives to Destroying Embryos for
        Stem Cell Retrieval   31

    Bioethics of Cloning   34

Stem Cell and Cloning Policies 36
    The Debate in the United States   36

    What Are Other Countries Doing?   39

The Future 40

Resources 42

Questions for Further Discussion or Research 44

Introduction

A woman with Parkinson's disease shuffles along a hallway with awkward, rigid movements and tremors. A man sits by a window in a wheelchair, ten years after an accident left him paralyzed from his neck down. Another man lies in a hospital bed after a heart attack has damaged part of his heart. What do these three individuals have in common? They have hope that stem cells may be able to help them to regenerate tissues in their bodies. **Regenerative medicine**, growing cells and tissues that can be used to replace or repair defective tissues and organs, is undergoing exciting changes as a result of recent advances in stem cell technology. Conditions such as these are just a few examples in which the treatment may drastically change in the future as our understanding of stem cells continues to improve. Some of the diseases that scientists believe stem cell technologies may one day play an important role in treating are listed in Table 1.

You are likely familiar with some of the controversy surrounding stem cells today, but did you know that some stem cell therapies have been around for decades? For example,

TABLE 1 Stem Cell–Based Therapies May Potentially Benefit Millions of People

Disease condition	Number of patients in the United States
Cardiovascular disease	58 million
Autoimmune diseases	30 million
Diabetes	16 million
Osteoporosis	10 million
Cancers (urinary bladder, prostate, ovarian, breast, brain, lung, and colorectal cancers; brain tumors)	8.2 million
Degenerative retinal disease	5.5 million
Phenylketonuria (PKU)	5.5 million
Severe combined immunodeficiency (SCID)	0.3 million
Sickle-cell disease	0.25 million
Neurodegenerative diseases (Alzheimer's and Parkinson's diseases)	0.15 million

Source: Table adapted from Stem Cells and the Future of Regenerative Medicine, www.nap.edu/catalog/10195.html.

bone marrow transplantation is a form of stem cell therapy. Bone marrow contains one kind of **adult-derived stem cells (ASCs),** the stem cells that regenerate tissues similar to those various, specialized tissues of the body in which they are found. During a bone marrow transplant, stem cells are transferred from a healthy donor to a needy recipient, where they then regenerate various blood cell types as needed. Obviously, then, some stem cells are already successfully being used in regenerative medicine today, and it looks like they continue to hold much hope for the future.

Scientists believe that another type of stem cell may have an even greater potential. The **embryonic stem cell (ESC)** is a type of stem cell retrieved from an early stage embryo. Unlike an ASC, which is limited in the type of tissue it can regenerate (e.g., a bone marrow cell only forms blood cells); an ESC can form all tissues of the body. Nevertheless, there are many hurdles yet to be overcome to make ESCs a viable treatment in regenerative medicine. While scientists can make many different cell types from ESCs, they still have much to learn about how these cells might behave if transplanted into a person. Additionally, much of the research needed to answer these questions has been slowed by political and ethical opposition to working with these cells. Traditionally, human ESCs have been retrieved from 5- to 7-day-old embryos, and retrieval of these cells destroys the embryos. Many people believe that ESC research should not be carried out if embryos are destroyed in the process.

While there is substantial political and ethical debate surrounding stem cells, scientists continue to make advances. Some scientists may avoid the moral debate by continuing the important search for new types of ASCs and new therapies utilizing them. Other scientists work within the framework of the current tense political climate. They continue to explore the possibilities of ESCs, including cells derived from controversial cloning techniques. In this booklet, we will examine what important discoveries scientists have made with stem cells and cloning, while concurrently being subjected to intense scrutiny.

Stem Cell Basics

WHAT IS A STEM CELL?

What makes stem cells such attractive candidates for repairing failing tissues and organs? A **stem cell** has two basic characteristics that make it unique from other cell types (Figure 1). The first is that it continues to grow and proliferate, maintaining a pool of cells just

FIGURE 1 Characteristics of a stem cell. Differentiation often occurs in various steps. Partly differentiated precursors, also known as progenitor cells, give rise to fully differentiated cells such as red blood cells or muscle cells. *Source:* Figure adapted from "Understanding Stem Cells," National Academies.

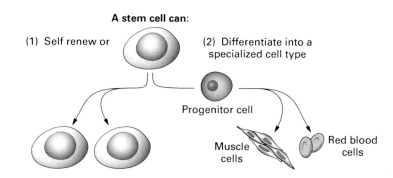

PATTERNS OF GENE EXPRESSION IN FIVE TYPES OF CELLS

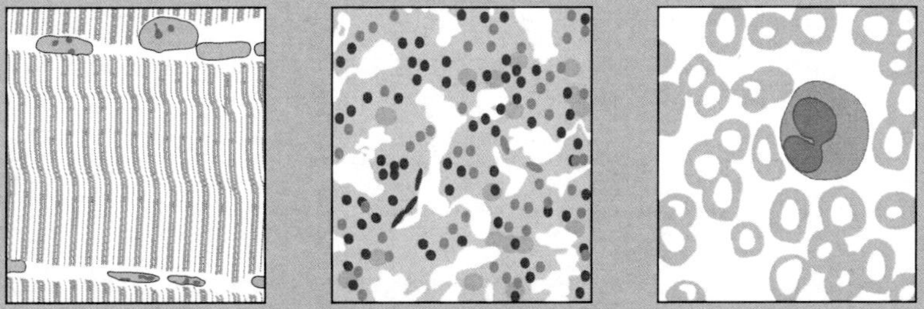

Genes for...	Muscle cells	Pancreas cells — Alpha cells	Pancreas cells — Beta cells	Blood cells — White cells	Blood cells — Red cells (immature)
Glycolysis enzymes	On	On	On	On	On
Muscle contraction proteins	On	Off	Off	Off	Off
Glucagon	Off	On	Off	Off	Off
Insulin	Off	Off	On	Off	Off
Hemoglobin	Off	Off	Off	Off	On

FIGURE 2 Unique "protein profiles" of five differentiated cell types. These differentiated cells have identical genes, but not all of the same genes are expressed in each cell type. If a particular gene is expressed in a specific cell type it is labeled as "on." Note that some genes are expressed in all cells (e.g., glycolysis enzymes involved in storing energy for a cell), while others are unique to one cell type (e.g., an oxygen transport protein, hemoglobin, is specific to red blood cells). *Source:* Figure adapted from Campbell, Reece, Mitchell, and Taylor, *Biology: Concepts and Connections*, Fourth Edition, p. 212.

like itself for possible future use (termed **self-renewal**). The second characteristic of a stem cell is that, given the correct signals, it can **differentiate** into a particular specialized cell type, such as a muscle or blood cell. When a stem cell divides, each of the daughter cells can either remain a stem cell or become a specialized cell. Let's further consider the concept of differentiation. Aside from gametes (egg and sperm cells), all the cells in an individual's body have the same DNA content and the same genes. What makes one of your muscle cells different from one of your blood cells? These cells are **differentiated,** or specialized to carry out a particular task (e.g., contraction in muscle cells or oxygen transport in red blood cells). While these two cell types contain the same genes, the specific set of genes that are being turned "on" or expressed are *not* identical. Since gene expression ultimately leads to protein expression, the "profiles" of proteins in specialized cells are different from each other (Figure 2). In fact, we all began life unspecialized. As our cells divided and began expressing a unique set of genes necessary for specific tasks, we each became a complex adult with over 200 differentiated cell types.

Some stem cells seem to have more abilities, or possibilities, than other stem cells; this flexibility is termed the **potency** of the stem cell. A stem cell that is **unipotent** can form only one differentiated cell type. A **multipotent** stem cell can form multiple different cells and tissue types. A **pluripotent** stem cell can form most or all of the 200 or more differentiated cell types in the adult body. A **totipotent** stem cell can form not only all adult body cell types, but also the specialized tissues needed for development of the embryo, such as the placenta.

WHERE DO STEM CELLS COME FROM?

At one time, it was thought that stem cells were only present in an embryo. Now we know that there are actually a lot of sources of stem cells, ranging from cells of the early embryo to the adult body. We can follow human development and examine the stages from which scientists have found stem cells (Figure 3). Traditionally, each of these stem cells has been categorized as either an embryonic stem cell (ESC) or adult-derived stem cell (ASC); but, as the research continues, the categories become less clear.

When a sperm cell fertilizes an egg cell, a single-celled embryo (**zygote**) forms. The zygote contains a complete set of genetic material (both the sperm and egg contribute half). This cell divides to become two cells, which divide to form four cells, and each of these cells divide to form an 8-celled embryo (three days after fertilization). Recently, scientists have demonstrated that cells of the 8-celled embryo can be removed from the embryo and grown in a laboratory dish to become ESCs.

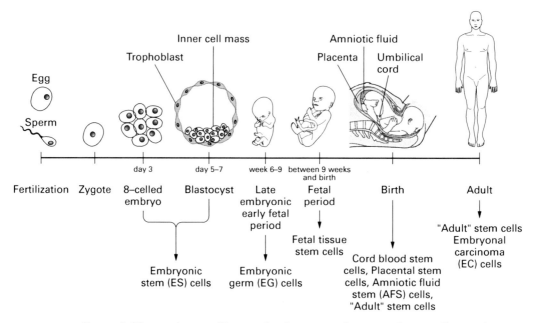

FIGURE 3 The continuum of human development and sources of stem cells at various stages. *Source:* Figure adapted from Prentice and Palladino, *Stem Cells and Cloning*, p. 5, Figure 2.

Around five to seven days into development, the human embryo consists of approximately 100 to 150 cells and resembles a hollow ball with some cells inside; this is termed a **blastocyst.** Since 1981, when two independent groups first established pluripotent stem cell lines from mouse blastocysts, scientists have focused on this stage in embryonic development as the key stage in development for human ESC retrieval. The outer layer of the blastocyst, called the **trophoblast** cells, consists of cells which develop to form parts of the placenta that are derived from the embryo. The cells inside the blastocyst are called the **inner cell mass**—these form all the cells of the baby. However, if these inner cells are removed and grown in a laboratory dish, these are now termed ESCs. Human ESCs were first retrieved from blastocysts in 1998 by Dr. James Thomson and his colleagues at the University of Wisconsin in Madison. Removal of the inner cells destroys the embryo. We will examine ESCs in much more detail later.

The largest potential source of ESCs is from the excess embryos produced by assisted reproductive technologies. Fertilization of an egg by sperm can occur either *in vivo* (inside a woman's body) or *in vitro* (outside of the body; e.g., in a test tube). *In vitro* **fertilization (IVF)** is a type of assisted reproductive technology that involves the removal of multiple eggs from a woman and fertilization *in vitro*. Resulting embryos are then implanted into the woman's uterus. Usually only a few of the embryos produced by the technique will be implanted, thus creating an excess of embryos in freezers around the country. These "leftover" embryos may be donated (with the couple's consent) for research. It is estimated that approximately 400,000 embryos are stored in clinics around the United States, although there is controversy about how many of these would actually be available for research purposes.

About the same time that human ESCs were isolated, Dr. John Gearhart and his team at Johns Hopkins University reported that they had isolated what they called **human embryonic germ (hEG) cells.** These cells are derived from the precursor cells that will become germ cells (egg or sperm) and were removed from early embryos that are developing into fetuses (around 6 to 9 weeks of development). When grown *in vitro,* the cells show many of the same characteristics as ESCs. That is, they seem to be pluripotent and able to form most or all of the tissues of the adult body.

As human development proceeds and various tissues start to form, the ESCs form **progenitor cells,** the partially specialized precursor cells that go on to form the specific differentiated tissues of the body. This is a gradual process, and the fetus continues to contain many multipotent stem cells (now termed **fetal tissue stem cells**) that form several different cell types.

The stem cells that are found in differentiated tissues of the body at *any* stage of development are commonly known as adult-derived stem cells (ASCs). The term "adult-derived" stem cell is actually not completely correct and can be confusing. What scientists often define as an ASC can be present before birth, at birth, or much later in life. Some other terms for ASCs that might be more inclusive and could be used interchangeably are: **somatic stem cell**, **tissue-specific stem cells** or **nonembryonic stem cells**. Nevertheless, we will use the term "adult-derived stem cell" to be consistent with the current nomenclature.

As we follow human development further, we find that the prenatal tissues necessary for supporting human development, such as the placenta and umbilical cord, are rich sources of stem cells. Multipotent stem cells derived from the umbilical cord—**cord**

blood–derived embryonic-like stem cells (CBEs)—appear to be less versatile than embryonic stem cells, but may have greater potential than the stem cells found in adult tissues. In early 2007, Dr. Anthony Atala from the Institute for Regenerative Medicine at Wake Forest University School of Medicine announced that he and colleagues had carefully characterized a group of pluripotent stem cells that seemed to behave halfway between embryonic stem cells and adult-derived stem cells. These cells can be found in the placenta and the amniotic fluid that surrounds the fetus in the womb and are termed **amniotic fluid–derived stem (AFS) cells.** As research continues with these prenatal tissues and the potential of their stem cells are realized, many experts believe they will become their own "category" of stem cells.

In thinking about the human body after birth, scientists have known for years that some tissues, such as bone marrow, contain ASCs, and they have been used in many successful clinical treatments. In recent years, ASCs have been identified in most differentiated tissues of the human body and have demonstrated variant potency. Although they are sparse and difficult to locate, they have been successfully isolated from tissues such as the brain, muscle, skin, pancreas, bone marrow, blood, and liver.

One last type of stem cell that is also part of the broad category of adult-derived stem cells actually comes from a tumor called a **teratoma** (if benign) or a **teratocarcinoma** (if malignant). (Scientists noticed that occasionally such tumors contained not just a disorganized mass of growing cells, as in most tumors, but also some differentiated tissues, such as a bit of bone, hair, or teeth! [Figure 4]). These types of tumors form when a germ cell (sperm or egg) spontaneously starts to grow and divide. The cells retrieved from these tumors, **embryonal carcinoma (EC) cells,** have the properties of stem cells. This led to research in which the EC cells were grown over a period of years and "tamed," so that they

FIGURE 4 An example of a benign cystic teratoma. Within this disorganized mass of tissue, teeth that have developed are obvious (*). *Source:* © CNRI/Photo Researchers

would not grow as disorganized tumor masses but instead form specific differentiated cell types such as nerves.

As you can see, there are various sources of stem cells. Keep in mind that each type has its advantages and disadvantages, and scientists continue to examine all possibilities. In general, the controversy that exists is summed up using the terms "embryonic" and "adult-derived" stem cells. Some scientists believe that pluripotent embryonic stem cells, which can form all tissues in the body, hold the greatest hope for treating degenerative diseases. However, there are many ethical issues to consider when using cells from early embryos (we will discuss this later). Adult-derived stem cells create less of an ethical dilemma but may not have the capabilities for treating disease such as is thought possible with embryonic stem cells. Both options present technical challenges that still must be overcome. Let's consider these two groups of cells in more detail.

Embryonic Stem Cells (ESCs)

Human ESCs have been touted as a "virtual fountain of youth" because of their potential to repair and rejuvenate any damaged tissue in the body. In theory, this should be possible because these are the cells that initially form all body tissues during development. As mentioned before, human ESCs were first isolated from the blastocyst stage of development. The procedure involves removing the inner cell mass of about 30 cells from inside the blastocyst (Figure 5). Once scientists isolate these cells, maintain them in a laboratory dish, and show that they have the properties of embryonic stem cells, we call this an **embryonic stem cell line.** These cell lines are continually growing, dividing, and crowding the laboratory dish. To keep them in a pluripotent, undifferentiated state, the cell line needs to be carefully cared for and replated to new dishes as they continue to proliferate. Millions of ESCs are derived from the original 30 cells after several months of replating.

Removing the inner cells destroys the embryo, and this is the main reason that research on human ESCs is so controversial. Scientists continue to search for ways to derive ESCs without damaging the embryo. Independent of how these cells are derived, some embryos have been, and will be destroyed during the scientific discovery process. However, if ESCs can perform all the wonders claimed for them in tissue regeneration, might it be acceptable that some human embryos are destroyed so that millions of lives can be spared? We'll first examine the scientific facts, and then we'll come back to this ethical question later.

CHARACTERISTICS OF EMBRYONIC STEM CELLS

Embryonic stem cells are pluripotent, meaning they can potentially form all tissues of the human body. This is what they do during normal embryological development; their "job description" is the initial formation of all the body tissues. Remember the characteristics of all stem cells: 1) continued growth and self-renewal, and 2) the ability to form differentiated tissues when given the correct signals.

We know that ESCs have an amazing ability for prolonged growth and self renewal, but why is this so? The ends of linear chromosomes are called **telomeres** and have been compared to the plastic caps on the end of shoelaces that prevent them from unraveling. With each cell division, the telomere progressively shortens. Once some critical amount of

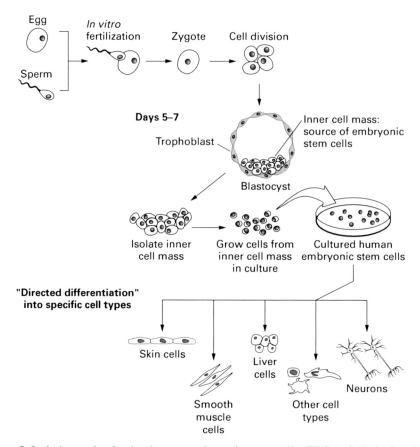

FIGURE 5 Isolating and culturing human embryonic stem cells (ESCs). Cells isolated from blastocysts can be grown in culture as a source of ECSs. Under the proper conditions, ESCs can be stimulated to differentiate into virtually all cell types in the body. *Source:* Figure adapted from Thieman and Palladino, *Introduction to Biotechnology*, p. 257, Figure 11.18.

the DNA at the telomere has been lost, a cell stops dividing and undergoes **senescence** (cell aging). Telomerase is an enzyme that counteracts this process. **Telomerase** repairs telomere length at the ends of chromosomes by adding DNA nucleotides to cap the telomere after each round of cell division. Telomerase is not active in typical differentiated cells, but scientists have observed that human ESCs express high levels of this enzyme. For example, after approximately 50 to 80 population doublings *in vitro,* typical differentiated cells will senesce. In contrast, several groups have shown that after 600 population doublings (more than three years in culture), human ESCs still continue to proliferate without apparent problems. Scientists believe that human ESCs can be grown in the laboratory in a pluripotent state indefinitely. In fact, some mouse cell lines have been growing for 30 years! However, these mouse ESCs and some human ESC lines have demonstrated that the longer a cell line is maintained in culture, the more genetic mutations it acquires. Eventually a buildup of small mutations "tip the scales," and the line is no longer useful. In sum, by preventing telomere shortening, telomerase activity is a major reason why ESCs can divide indefinitely.

As for the ability of ESCs to differentiate, scientists have been able to directly differentiate these cells into many unique cell types (see Figure 5). A few of the cell types into which ESCs have already been made are: heart, nerve, cartilage-forming, immune, skin, bone, adipocyte (fat), pancreatic, skeletal muscle, smooth muscle, and blood vessel cells. Research continues in this area of **directed differentiation** to discover the correct signals needed to produce the various cell types of the human body (think of this as tweaking a recipe so that it is just perfect). The signals that stimulate differentiation of stem cells include hormones, molecules called growth factors, and small proteins. Additionally, once a specific cell type is obtained, the cells need to undergo testing to be sure they actually function as expected.

TESTS FOR PLURIPOTENCY, THE ABILITY TO FORM ANY BODY TISSUE

What is the basis for the claim that an ESC can form any adult tissue? It is based on several different types of scientific studies. First is the simple fact that, under normal developmental conditions in an intact embryo and when left alone to do their job, ESCs will form all the tissues of our bodies.

Scientists have devised several tests for pluripotency. For ethical reasons, not all of these have been performed with human ESCs, but rigorous experimentation with mouse ESCs have demonstrated the pluripotency of these cells. In all tests, scientists look to see if the ESCs differentiate into cells that represent each of the three major layers of an early embryo (Figure 6). During embryological development, the inner cells first form three semi-specialized layers, the three **primary germ layers,** which then form the specific tissues in the body. These tissues are committed to a developmental pathway at this point. The outer layer, **ectoderm**, gives rise to skin, brain, and nerves. The middle layer, or **mesoderm,** forms blood, heart, bone, kidney, muscle, and cartilage. The innermost layer, the **endoderm,** develops into the lung, liver, and digestive system. It has been well demonstrated *in vitro* that human ESCs can differentiate into numerous cells types that represent all three germ layers. It is necessary, of course, to also demonstrate their abilities in a living system (*in vivo*), where they are exposed to the normal signals that cells experience.

In vivo tests to demonstrate pluripotency with human ESCs are limited. One that has been performed with several human cell lines tests their ability to form all three germ layers upon injection into mice. Injecting foreign cells into a normal mouse would cause a major immune system response, and these cells would be rejected by the host mouse. For this reason, these tests are performed with **immunodeficient** mice—that is, mice lacking a functional immune system. Such mice need to be kept in special sterile environments so that they do not get an infection; even a common cold might be deadly. Their lack of an immune system means that they will not reject the transplanted human cells. When embryonic stem cells are placed in these immunodeficient mice, tumors form. These tumors are similar to teratomas, with some of the ESCs differentiating into specialized cell types and

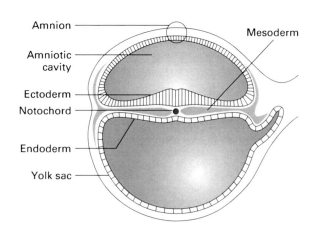

FIGURE 6 The gastrula stage of human development. The three primary germ layers are formed (i.e., the ectoderm, the mesoderm, and the endoderm).
Source: Figure adapted from http://www.britannica.com/eb/art-2921/Development-of-the-human-embryo?articleTypeId=1.

tissues. The tumors are examined closely to see what types of specialized cells grow. For example, if ESCs that are injected into the immunodeficient mice form nerve cells, heart cells, and intestinal cells, this indicates that they should be pluripotent. Why? These three cell types each originate from a different primary germ layer. Although they may not form every single tissue type, experiments such as these have demonstrated that human ESCs have the ability.

Most scientists agree that the "gold standard" for demonstrating pluripotency *in vivo* would be to inject human ESCs into a developing embryo at the blastocyst stage (the same stage used to isolate embryonic stem cells). The resulting developing fetus would then be examined to determine if the injected ESCs contributed to development of all three germ layers. Experiments such as these have been performed with mouse cells—putting ESCs from one type of mouse into the embryo of another type of mouse (Figure 7). For example, if embryonic stem cells from a black mouse were put into an embryo for a white mouse, then the coat color of the mouse once born should be both white and black. (Other genetic markers can be tracked to see what tissues the injected stem cells helped to form internally, too.) Using these types of experiments, mouse ESCs have been able to help construct most or all of the body tissues of the mouse born.

Rather than using human embryos as hosts, which has major ethical implications, some scientists would like to perform similar experiments to test the pluripotency of human ESCs in which the human embryonic cells are injected into *mouse* blastocysts. The resulting fetus would similarly be examined for human cell contribution to various mouse tissues. However, the creation of embryos which are part human and part mouse, termed **human-mouse chimeras,** also has been strongly opposed on ethical grounds. Many people worry that this type of research is a "slippery slope" and that these experiments might lead to allowing adult mice, with human tissues, to be born. As eloquently stated by Nancy L. Jones, from the Center for Bioethics and Human Dignity:

> *Until such an experiment is actually conducted, there is no way of knowing if human stem cells could even produce tissue in a mouse, if such tissue would grow normally and function, or if all types of animal tissue could be converted to human tissue (the intriguing question emerges here of whether*

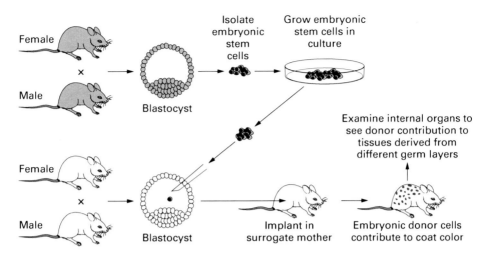

FIGURE 7 Evidence that an embryonic stem cell is pluripotent and can contribute to all tissues. Mouse ESCs from one mouse (black) can be transferred to a blastocyst (produced from white mice). The resulting mouse will be a chimeric mouse. Various tissues, representing the three germ layers, will have formed from both the white and black stem cells.
Source: Figure adapted from Prentice and Palladino, *Stem Cells and Cloning*, p. 8, Figure 5.

we would feel differently about human stem cells that contributed to the liver, rather than, say, to the brain). In the face of such ambiguity, the overarching ethical question is: Should we even begin these types of experiments?

For now, scientists rely on the experiments that have been performed with human ESCs to demonstrate pluripotency and infer much from the numerous *in vivo* studies that have been performed with mouse ESCs. Interestingly, the nature of pluripotency remains a mystery. For example, what controls pluripotency? A hot area of stem cell research is to find a gene expression "profile" of a pluripotent stem cell in order to better understand the signals that control it.

WHAT ARE THE POSSIBLE USES FOR EMBRYONIC STEM CELLS?

Many of the hopes for ESCs are based upon the past two decades of basic research with mouse ESCs. Medical research was transformed when scientists learned how to delete, add, or change genes in mouse ESCs. In fact, the 2007 Nobel Prize in Medicine was awarded to three scientists for their pioneering work in this area. By genetically engineering mouse ESCs, scientists learned to model many different human genetic diseases in mice, such as hemophilia and cystic fibrosis. Embryonic stem cells from many other mammals, such as cows and monkeys, have also provided valuable information about how these cells behave. The knowledge gained by working with various ESCs has led to important animal models for most major genetic diseases, and many useful therapies have followed. Research in the area of human ESCs is relatively new because these cells were not isolated until 1998. Let's explore some of the possible uses for human ESCs.

The hopes for human ESCs are enormous (Figure 8); and, despite what is most publicized in the media, not all of these hopes focus on direct use in patients. For example, many scientists see these types of cells as a basic science research tool to learn more about normal embryonic development. That is, what differentiation factors trigger development of progenitor cells into specialized tissues? How is pluripotency defined by a cell? What genes are specifically turned "on" or "off" in a pluripotent cell compared to a differentiated cell? Although scientists have learned much about human development using various animal models, there are significant differences between animals and humans. In fact, researchers are already finding that the signals that keep mouse ESCs and human ESCS in an undifferentiated state *in vitro* differ.

Another less obvious use for human ESCs is the ability to closely study specific genetic abnormalities. For diseases in which we know the genetic basis, scientists have the ability to add, delete, or alter genes in human ESC lines to model the human disorder at the cellular level. However, for more complex genetic diseases that require unknown causes for onset, scientists could use the genetic material from a person with that complex disease to make a new, unique ESC line (this is called therapeutic cloning, and we will discuss it later). What types of questions could be answered with these "diseased" ESCs? For example, when and why does a pancreatic cell stop producing insulin in a diabetic individual? Scientists could examine the process of pancreatic differentiation in ESCs that are normal and diabetic and compare them under different environmental conditions. In this way, scientists will learn much about how and why cells become "sick."

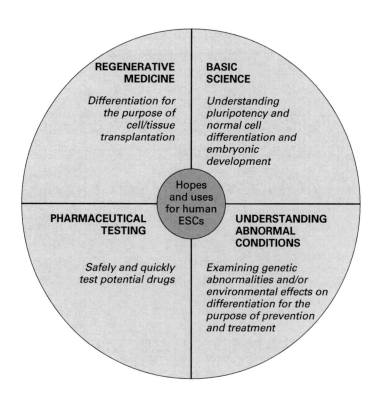

FIGURE 8 A summary of the hopes and uses for human embryonic stem cells.

An important application of studies of human ESCs would be in the field of pharmaceutical testing. Consider this example: a scientist develops a drug for heart cells and needs to test the drug on human heart cells before going into human clinical trials. Because it is not possible to maintain human heart cell lines, the scientist must use animal models. Because animals and humans differ, the information gained about the drug's effectiveness or toxicity may not necessarily apply to humans. However, with the ability to differentiate human ESCs into heart cells, scientists can make an appropriate cell line at the time a drug is ready for testing. This type of application could lead to the faster development of safer and more effective drugs.

Of course, the most publicized hope for human ESCs is their possible use in regenerative medicine, in which failing body parts would be replaced with ESC-derived tissue. Unlike many therapies that exist today, this could, in theory, provide an unlimited supply of tissue. Some diseases are caused by the loss or dysfunction of only one or a few cell types. For example, insulin-producing pancreatic cells do not function in diabetes, and dopamine producing neurons are lost in Parkinson's disease. Scientists envision injecting a few ESC-derived pancreatic cells or neurons into the area in which they are needed and allowing the body to direct the regeneration. Others envision growing more patterned tissues and even whole organs that could repair larger failing parts. Additionally, with a process called therapeutic cloning, which we will discuss in detail later, these cells could theoretically be customized to an individual to avoid immune rejection. For example, if you had a heart attack, physicians could take some of your DNA from, say, a skin cell and make new ESCs that contain your genetic material. Once these cells were differentiated into heart cells, they could be used to repair your damaged heart!

So, human ESCs hold much promise for regenerative medicine, but they are not ready for use in humans. However, various experiments with ESCs demonstrate their potential. For example, mouse ESCs that were differentiated to endoderm-like precursor cells and injected into mouse livers were able to cure these mice of hemophilia. Human ESCs that were differentiated into heart cells and injected into pig hearts lacking normal electrical signals were able to reestablish heart function. Human ESCs that were differentiated into dopamine-producing neurons were able to significantly reduce symptoms in a rat model of Parkinson's disease when injected into rat brains. There are many more examples of promising uses for ESCs, so why are these applications so slow to reach human clinical trials?

WHAT ARE SOME OF THE PROBLEMS WITH EMBRYONIC STEM CELLS?

Although they are relatively easy to identify, expand, and grow, there are many reasons why human ESCs are not likely to be used in the immediate future in regenerative medicine. In addition to ethical and political barriers that we will discuss later, safety is a major concern. Controlling which cell types ESCs will differentiate into when injected into the body is a major barrier. For example, when the rats with Parkinson's disease were injected with human ESC derived neurons, many of the rats developed tumors. Why? When pluripotent ESCs are injected into animals, they form tumors called teratomas. Although these human ESCs were differentiated into neurons, it is likely that not all were fully differentiated. For these reasons, many scientists believe it will be best to differentiate ESCs into maturing cell types that would be injected into the body instead of injecting ESCs that could potentially differentiate into unwanted cell types.

Another hurdle to overcome for clinical therapy is the immune system of the host. Generally, when foreign cells are injected into a host, the immune system of the host attacks and destroys these cells. It is not yet clear from animal studies if this would be a major problem for humans, although many scientists think it is likely to be. As with other transplants, a patient could take immunosuppressive drugs to minimize this immune response. Imagining much further into the future of medicine, we could envision ESC lines that are genetically altered to avoid detection by any patient's immune system or ESC lines that are customized for each person. If we can overcome the issues of safety and immune rejection, what will be the long-term fate of these injected cells in a person? Once introduced into the body at a specific location, there is no guarantee that the cells would remain at that location. There is still much to be learned from animal studies that will affect our ability to safely use human ESCs in humans. Nonetheless, Geron, a company working to cure spinal cord injuries, surprised many when they announced that they were seeking FDA approval to begin injecting ESC-derived cells into humans. Geron was the first company to make such an announcement and, as of March 2008, still awaits FDA approval.

Adult-Derived Stem Cells (ASCs)

As mentioned earlier, the label of "adult-derived" stem cells is a bit of a misnomer in that these cells can be found in any differentiated tissues at any stage of development. So, what we call "adult-derived stem cells" is a broad category of cells, each with unique potential. For now, we can think of ASCs as nonembryonic stem cells. Many differentiated tissues have been found to have ASCs. They have been identified in the skin, bone marrow, fat, and muscle, to name just a few!

CHARACTERISTICS OF ADULT-DERIVED STEM CELLS

Traditionally, we think about ASCs being different from ESCs for two major reasons: 1) they can not grow indefinitely, and 2) they are limited in their ability to differentiate. Adult-derived stem cells can be very challenging to grow in the lab because they do not divide often or easily. For example, they may only double one time *in vitro* before dying. Compounding this issue, they are often present in tissues in small quantities and can be difficult to identify, purify, and isolate. For instance, in mouse bone marrow, only 1 in approximately 10,000 cells is a stem cell. Even if a few ASCs are isolated, being able to grow large numbers of them in the lab can be difficult. The reason ASCs are often labeled as "limited" is because of their potential to regenerate a narrow range of damaged tissues. That is, skin stem cells only give rise to skin. Blood stem cells only give rise to various blood cells. Right? But what if a blood stem cell from the bone marrow were put into a different environment, such as a liver? Would environmental cues from the liver affect what that blood stem cell became? Think about it as analogy with yourself...would you be the same person you are today if at birth you were raised in a different location with a different family? Probably not! Your environment (culture and experiences) play a major role in your development. If you grew up in a completely different place, you might speak a different language, have a different hairstyle, have different hobbies, and a different career goal. Similarly, scientists have placed blood stem cells (that normally make blood cells) into host mouse livers and observed that some of the blood stem cells became liver cells.

Observations such as these support the hypothesis that some ASCs may be **plastic,** or changeable, in response to regenerative signals. (Plasticity is sometimes referred to as **transdifferentiation** if the differentiation event causes the cell to cross the germ layer barrier, such as from mesoderm to endoderm; Figure 9). Examining bone marrow transplants between people of different genders was one of the initial observations that suggested that plasticity might be possible. For example, when bone marrow from a male (in which all the cells contained a Y chromosome) were transplanted into a female host (in which all the cells contain only X chromosomes and no Y chromosomes), some cells containing Y chromosomes were found in the liver; this result was unexpected because it was thought that bone marrow cells only gave rise to blood cells. The Y-containing cells could only have been derived from the bone marrow transplant. Similar observations showed

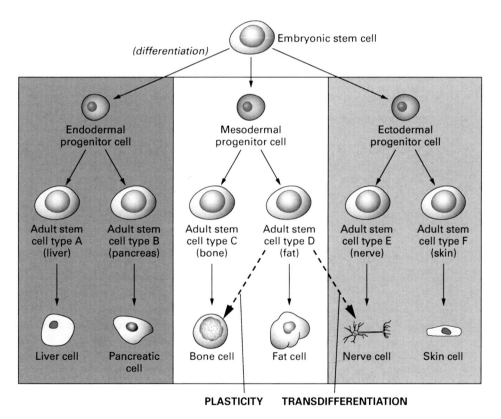

FIGURE 9 The concept of plasticity overlaid upon a more traditional approach to thinking about adult stem cell (ASC) differentiation. The ability of ASCs to become other cell types is referred to as plasticity. If the change crosses a germ line barrier, this is a specific type of plasticity known as transdifferentiation.

that Y-containing cells were found in other unexpected tissues, such as the heart and brain. These experiments generated much hope for ASCs and suggested that they might have as much potential as ESCs. However, alternative explanations exist for these results. For instance, one group of scientists demonstrated with mouse experiments that transplanted bone marrow cells that appeared in various host tissues were actually due to cell-to-cell fusion (i.e., two cells fuse to become one large cell with two sets of genetic material).

There is now much debate as to whether plasticity is a real phenomenon *in vivo*. Many scientists think that if this type of event can occur with bone marrow cells *in vivo,* it happens so rarely that it is not likely to be useful in terms of treating various organ failures with bone marrow transplants. Interestingly, we are seeing many new reports now that various adult stem cells grown *in vitro* demonstrate plasticity. For example, skin stem cells can become bone, muscle, or fat, and fat stem cells can become bone, cartilage, and nerve. Can you imagine the potential of the stem cells in the thousands of pounds of skin and fat being discarded from procedures such as liposuction every year in the United States? We are likely to see many new therapies being developed based upon the *in vitro* plasticity of these cells in the future.

ADULT-DERIVED STEM CELLS ARE ALREADY IN USE IN REGENERATIVE MEDICINE

So while there is still a lot unknown, many scientists think that most of the body's own tissue repair mechanisms *in vivo* come from ASCs that reside in *each* tissue. That is, if your liver needs repair, liver stem cells—not circulating multipotent stem cells or ASCs from other tissues with high plasticity—repair it. While individual ASCs may be limited in their potency, clinicians have made amazing progress in the area of regenerative medicine by utilizing these cells. Scientist can take one of two approaches when thinking about the regenerative potential of ASCs in each tissue. The first approach would be to expand a patient's own tissue-specific stem cells in the lab and then transplant numerous derived cells or organized tissues into the patient. The second approach would be to get the resident stem cells to become more active after injury by stimulation with a pharmaceutical.

To consider both approaches, let's look at the heart as an example (Figure 10). Because the heart is not very good at regenerating heart muscle after injury, it was assumed for a long time that the heart did not have ASCs. A normal repair mechanism after a heart attack is the replacement of dead muscle cells with noncontractile scar tissue. However, in 2004 and 2005, various research groups identified these elusive cardiac stem cells. Let's first consider the transplant approach to repair the heart. Physicians initially would isolate a small amount of heart tissue (about the size of a grain of rice) through a thin tube from a patient undergoing a cardiac biopsy. Some of the cells in this tissue would be heart ASCs. After approximately three weeks of growth in the lab, there would be enough stem cells to transplant the cells into an injured heart. If the cells went back to the same patient from which they were originally isolated, this is termed an **autologous transplant.** A benefit to this type of transplant is that there will not be immune rejection because the cells are from the same person. (This is in contrast to **allogenic transplants,** in which cells are transplanted from one person to another.) While this type of cardiac stem cell transplant therapy

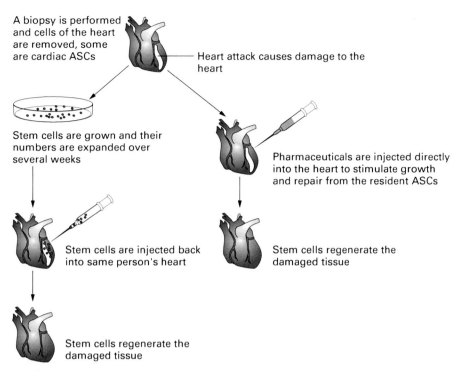

FIGURE 10 Two possible approaches utilizing adult-derived stem cells (ASCs) of the heart to regenerate damage after a heart attack. *Source:* Figure adapted from Thieman and Palladino, *Introduction to Biotechnology*, p. 259, Figure 11.19.

is not being done with humans yet, studies in mice and pigs show promising results; the injected cells migrate to the damaged zones of the heart and repair damaged tissue.

Transplantation may be a solution for some patients, but it also has its drawbacks. For example, a patient in critical condition who has just had a heart attack may not be able to wait three weeks for treatment. Thus, a pharmaceutical approach to treating heart attacks may be the preferred treatment for other patients. With this approach, researchers try to develop chemicals that might cause the few residing ASCs to become more active or recruit other stem cells from different places in the body. For example, in one study designed to activate the resident stem cells of the heart, scientists injected various drugs directly into the hearts of dogs with a heart injury. The dead tissue was significantly repaired with new heart cells that contracted and improved overall function. Other scientists are searching for chemicals that cause the process of **dedifferentiation,** when a body cell (such as the injured heart cell or a cell nearby) might lose its specialized, differentiated characteristics and become a stem cell (Figure 11). While this is a new area of research for therapeutic purposes, dedifferentiation has been well studied in the tail regeneration of some amphibians, such as salamanders. When the tail is cut or pulled off, scientists have observed that various mature cells, such as muscle cells, dedifferentiate and eventually give rise to a new limb, including spinal cord, muscle, bone, and skin.

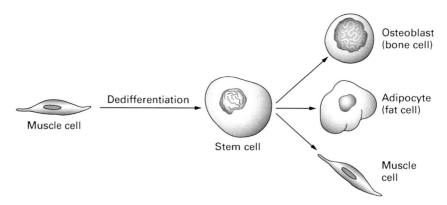

FIGURE 11 Dedifferentiation of a specialized body cell (such as a heart muscle cell) might cause it to become a stem cell that can give rise to new heart muscle cells or other cell types. *Source:* Figure adapted from "Future of Stem Cells," SCIAM, 2005.

Transplanting ASCs holds much promise in the future...but wouldn't it be amazing if entire organs could be grown from few stem cells? This has already been done! As you likely know, there is quite a shortage of organs available for transplants; in 2007, the U.S. government estimated that 94,000 people were waiting for organs, and that number will continue to increase. Whereas allogenic transplants save lives every day, there are serious complications with immune rejection. What if your *own* cells could be used to grow a new organ that could then be transplanted back into you? This was once the material of science fiction novels but now is reality. In 2006, Dr. Anthony Atala and colleagues at Wake Forest University School of Medicine announced that they had done just this for seven patients who were born with defective bladders (Figure 12). Whereas most cells of the bladder will only grow and divide for a few days, the team removed from each patient a small amount (dime sized) of the bladder that was present and that included ASCs. They grew the cells on a hollow, biodegradable scaffold, and, after many weeks of growth, transplanted the artificial bladder back into the patient. After seven years, the patients still have much improved or normal bladder function without ill effects. These artificial bladders marked the beginning of an era for growing artificial organs for autologous transplant using ASCs. Dr. Atala's group is working on more than 20 different organs using a similar technique.

What scientists are learning about these many different kinds of ASCs and how this knowledge is being utilized in medicine is changing daily. Sometimes, scientists learn the basic science behind stem cells in the lab and hope to translate this to a clinical use. Scientists at Duke University Medical Center and Pratt School of Engineering have found that fat stem cells isolated from human liposuction procedures could be efficiently coaxed to become cartilage cells *in vitro*. Cartilage injury is very difficult for the body to repair on its own, and current methods with cartilage transplants don't work well. Know-

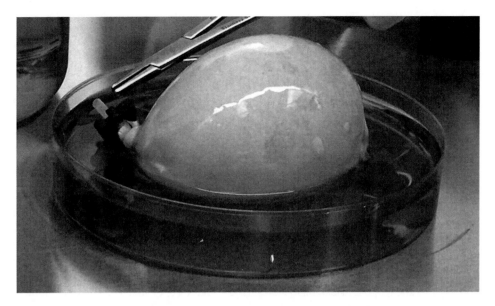

FIGURE 12 A real application of adult-derived stem cells—growing human organs. A biodegradable scaffold that is molded to the shape of a bladder is seeded with a patient's own bladder stem cells. After weeks of growth, these artificial bladders are transplanted into the patient. *Source:* Brian Walker/AP Photo, http://news.nationalgeographic.com/news/2006/04/0404_060404_bladders_2.html.

ing how to convert fat cells into cartilage cells puts scientists in a good place to develop clinical stem cell therapies based upon their findings. Yet sometimes a reverse approach is used. For example, there is some improvement in cardiac function when a person's own bone marrow stem cells are infused into his/her heart after a heart attack. So how does it work? Scientists don't completely understand how the stem cells may cause improvement; it is possible that they produce signals that cause other cells to respond and repair the heart. So whether scientists move toward a stem cell therapy from the basic science or develop the therapy first, this is certainly an exciting time in regenerative medicine.

A Multipronged Approach to Curing Diseases

In general, scientists want to pick the most efficient, safest, and quickest way to cure a disease. For this reason, we see each disease being approached from many angles (e.g., pharmaceuticals, gene therapy, surgery)—not just with stem cells. Scientists work together in curing a disease with a multipronged approach. Not surprisingly, when we examine how stem cells are affecting basic research on diseases, we can find researchers using ESCs, fetal stem cells, and various types of ASCs. To make an example of the many approaches within the stem cell field that are being used to solve a common problem, let's examine the neurodegenerative disease, Parkinson's disease. You are likely somewhat familiar with this disease, since Michael J. Fox has made his disease and foundation public.

Parkinson's disease (PD) is a debilitating disease affecting approximately 1.5 million older Americans (the age of onset is usually around 65 years). The disease is caused by a loss of specific neurons, or nerve cells, in the brain. These neurons normally produce a chemical messenger called dopamine that is responsible for the smooth, coordinated movement of muscles. When approximately 80% of the dopamine- producing (DA) neurons are lost, symptoms appear; these include slowness of movement, stiffness, shaking, and difficulty balancing. Currently there is no cure, but pharmaceuticals can help lessen the symptoms. A common pharmaceutical that is used for treatment, called levodopa or L-dopa, has been in use for decades. While levodopa helps to elevate levels of dopamine in the brain, patients respond less to it over time. Other pharmaceuticals are difficult to develop due to the blood-brain barrier that prevents many substances, including drugs, from entering the brain. Stem cells are now offering hope for new treatments and maybe even a cure for PD. However, predicting what type of stem cell will prove most useful years from now would require a crystal ball. We are at the cutting edge with stem cells, so let's just examine some of the fascinating studies to date that relate to stem cells and PD (Table 2)

Some studies have utilized ESCs as the starting cell type for trying to ameliorate PD in animal models. For instance, monkey ESCs have been used to generate neurons. With the addition of a particular growth promoting factor, DA neurons were effectively produced. When transplanted into the brains of monkey models of PD, the DA neurons functioned and diminished symptoms of the disease. However, only 1 to 3% of the transplanted cells survived, indicating that there is room for much improvement before this type of technique can be thought of as a therapy for PD. Better survival and incorporation of the DA neurons would be necessary. Similar experiments have been done in which human ESCs generated DA neurons. When transplanted into the brains of rat models of PD, the treatment improved muscle coordination such that the rats were almost normal. A pitfall to these, and other similar experiments with mice, was highlighted by the multiple tumors found in the brains of these animals. Better control over differentiation to lessen the odds of tumor development needs to be possible before trials with humans can take place. (Recall from the previous discussion that undifferentiated ESCs injected into an animal cause tumors.) In line with this, new methods have been developed with mouse cells to define and identify a pure population of completely differentiated neural cells (derived from ESCs). Applied to transplant studies, the methods are proving useful in removing the unwanted side effect of tumor formation.

Other studies similarly designed to improve PD in animal models have used fetal stem cells rather than ESCs as the initial cell type. Specifically, **human neural progenitor cells (hNPCs)** obtained from fetal tissues at 10 to 15 weeks' gestation have been examined. In one study, these cells were genetically modified to produce a survival factor for DA neurons and were then transplanted into the brains of rat or monkey models of PD. These cells continued to produce the survival factor, which led to DA neuron generation and subsequent dopamine production. Physical improvements that lasted for several months were seen in the animals. Nonetheless, these promising studies indicate that this is not a long-lasting cure and that controlling the levels of the survival factor must still be refined.

TABLE 2. A Summary of Some Stem Cell Experiments Related to the Neurodegenerative Parkinson's Disease (PD).

Source cells	Differentiated cell type	Host animal receiving brain transplant	Results
Monkey ESCs	Dopamine-producing neurons	Monkey model of PD	Diminished PD symptoms. Low survival rate of the transplanted cells
Human ESC	Dopamine-producing neurons	Rat model of PD	Significantly improved muscle coordination. Tumor formation in brains
Human neural progenitor cells (hNPCs) from fetal tissue engineered to express a "survival factor"	N/A	Rat and monkey models of PD	Improved symptoms of PD. New dopamine-producing neurons generated. Effects were not long lasting
Adult human brain biopsy cells	Neural progenitor cells	Mouse	New neurons generated
Mouse or human neural ASCs	N/A	Mouse model of the related disease, Sandhoff's disease	Increased life span. Delayed loss of motor function. No tumors
Human ESCs	Neural progenitor cells	Mouse model of the related disease, Sandhoff's disease	Increased life span. Delayed loss of motor function. No tumors

N/A = not applicable.

Adult brain tissue has also been used in studies designed to correct PD. For example, when brain cells from patients undergoing surgery were grown in the lab in the presence of a specific growth-promoting factor, scientists noticed that neural progenitor cells developed. (Remember, progenitor cells are a bit further down the pathway of differentiation compared to stem cells.) When these progenitor cells were transplanted into mouse brains, scientists observed development of new neurons and incorporation of these cells into various regions of the brain. If these strategies can be applied to correct PD in humans, there are advantages to beginning with adult cells as the starting material (e.g., less chance of tumor development, a person's own cells could be used for autologous transplantation, and fewer ethical and political issues compared with using embryonic and fetal tissue).

A recent study with **Sandhoff's disease,** a neurodegenerative disease related to PD, has offered new hope for PD patients. Scientists transplanted adult neural stem cells from mice or humans or fully differentiated human neurons derived from ESCs into the brain of a mouse model of Sandhoff's disease. All types of stem cells prolonged the lifespan, delayed the loss of motor function, and, importantly, did not invoke immune rejection or tumor formation. An important first discovery was also made—the mouse neural cells replaced damaged neurons and were able to transmit impulses. Lastly, these studies clearly illustrate the multipronged approach that will eventually include stem cells. When these mice were given a specific pharmaceutical, in addition to the stem cells, their lifespan doubled. In fact, the two treatments together demonstrate *more* than an additive result (i.e., the results were better than the sum of the individual effects of each treatment), and this is termed a **synergistic effect.** In sum, while stem cells are certainly a new frontier in the battle against diseases like PD, they will not be the only frontier. Without doubt, stem cell treatments will be routinely combined with other important therapies.

Cloning

When people think about cloning, images are conjured up of Elvis returning from the past or armies of identical soldiers in science fiction movies. However, cloning does not simply apply to an entire organism. Molecules, such as DNA, or individual cells can be cloned—cloning processes make multiple copies of the original DNA molecule or cell. These processes have been done in the lab for decades, and many of the modern medical breakthroughs have been based on cloning DNA and cells. When an entire organism is cloned, the complete adult animal is not immediately produced. The clone starts as a one-celled embryo and must still go through an embryonic stage, grow, and develop. Frogs were first cloned in the 1960s, but it was not until 1996 that the first mammal, Dolly the sheep, was cloned using an adult sheep cell nucleus. Dolly was born in 1997.

WHAT IS CLONING?

To understand cloning, let's first begin by thinking about sexual reproduction—the union of a sperm and an egg. During sexual reproduction, the sperm and the egg each bring half of the chromosomes for the new individual, so that the one-celled embryo contains a full set of chromosomes (half from mom and half from dad). This process can happen *in vivo,* inside a female's reproductive tract or *in vitro.* The latter would be termed *in vitro* fertilization, and the embryo would be placed into the uterus after several days to go on to form a normal baby. (As we discussed previously, if the *in vitro* embryo was grown to the blastocyst stage and then dissociated instead of being implanted into a uterus, an ESC line could be generated.) In the end, sexual reproduction produces offspring that are *not* identical to either parent.

So how does cloning differ from sexual reproduction? Unlike sexual reproduction, a sperm cell is not needed. The process of cloning requires two main ingredients: 1) the genetic material from a **somatic cell** (a body cell other than a sperm or egg) of the organism to be cloned, and 2) an egg (Figure 13). *In vitro,* the nucleus (containing the genetic material) is first removed from the egg, a process termed **enucleation.** Next, the nucleus containing a complete set of genetic material is taken from a somatic cell and inserted into the enucleated egg. This cloning process creates a one-celled embryo. Contrary to sexual reproduction, in which the sperm and egg each donate half the chromosomes, the somatic cell donates all of the chromosomes needed to make an individual. Terms used to describe this process are **somatic cell nuclear transfer (SCNT)** or **nuclear transplantation.** At this stage, the embryo is then stimulated to divide and grow with the help of electric shock or chemicals. Once the embryo has developed for several days *in vitro,* the embryo must be placed in the uterus of a surrogate mother to complete development and undergo normal gestation until birth. The organism produced by this method, a **reproductive clone,** will be genetically identical to the somatic donor. In contrast, if the embryo is never placed into a uterus, it can be dissociated and an ESC line can be generated. These stem cells can be differentiated into various tissues for the purpose of treating patients. This is termed **therapeutic cloning.**

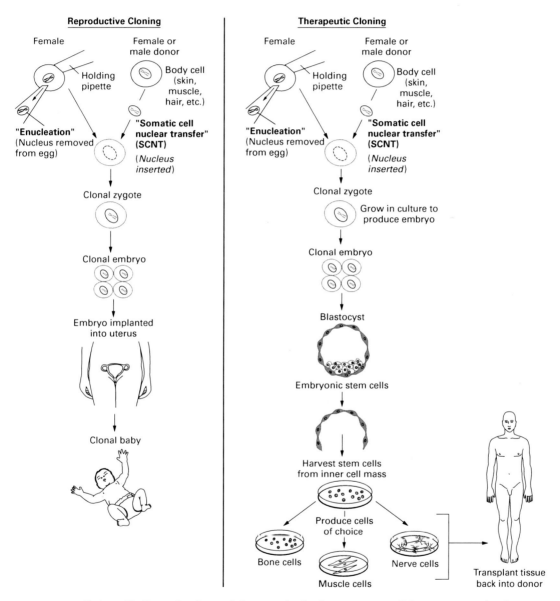

FIGURE 13 Reproductive and therapeutic cloning use many of the same procedural steps. *Source:* Figure adapted from Thieman and Palladino, *Introduction to Biotechnology,* p. 261, Figure 11.20.

PURPOSES OF CLONING—WHY DO WE WANT TO CLONE?

So why would we want to make a clone? For agricultural uses, there are many reasons why having cloned animals could be an advantage. Selectively breeding the best livestock—for example, crossing the best bull with the best cow—can produce champion animals, such as good milk or wool producers; but, when the champion animal is bred, there is no guarantee that its offspring will have all of the same characteristics because only half of the genetic makeup of the offspring will be from that champion animal. However, if we could produce a clone of that animal, then the cloned offspring would have the same genetic makeup of the champion. It is even possible to genetically engineer animals so that their milk contains medically valuable human proteins, such as insulin, or so that their organs can be used for transplantation and not be rejected by humans transplant patients (transplanting animal organs into humans is termed **xenotransplantation**). For these types of animals, cloning could ensure that there are many identical animals available. Cloning could also be useful in producing **animal models of disease**—animals that have a specific disease—so that the medical condition can be studied in attempts to find treatments and cures. Interestingly, another use that was not commercially successful was attempted by a company in California. For $50,000 this company would clone your pet cat. Apparently there was not enough of a demand, and the company closed its doors!

For human cloning there are two purposes proposed. One is **reproductive cloning,** sometimes called **live-birth cloning.** After the SCNT process, the cloned human embryo would be implanted into the uterus and would develop until birth. Proponents of this purpose for human cloning talk about using it to reproduce a child who has died or to help infertile couples to have a child. The other purpose for human cloning has been called **therapeutic cloning** or **research cloning**. The idea here is to use SCNT to produce a cloned embryo that is identical to a patient who has a degenerative disease. Yet instead of implanting the embryo in a uterus, the embryo is dissociated and ESCs are retrieved, in hopes of treating the patient. In theory, the cells derived from the ESCs, such as liver or pancreatic cells *etc.*, would be a genetic match for the patient and would not cause transplant rejection.

REPRODUCTIVE CLONING

Reproductive cloning—cloning to produce a live birth—has been accomplished with several other mammals thus far. Since Dolly was cloned, other mammals that have been cloned and brought to birth are mice, goats, pigs, cattle, rabbits, dogs, and cats. Cloning primates (e.g., monkeys and humans) has been more challenging. The embryos produced by SCNT in primates have not developed past an early stage. This points out one of the biggest problems in cloning—most clones do *not* survive.

Cloning, as it turns out, is a very inefficient process. It took 277 tries to get *one* Dolly. Each attempt involved creating a cloned embryo by SCNT. Mammary gland cells from the udder of a 6-year-old sheep were used as the somatic donor nuclei, fusing these cells with an enucleated egg to produce the one-cell clones. Only about 10% of these single-celled cloned embryos typically develop to the blastocyst stage (in Dolly's case, only 29 of the

277 cloned embryos). Those that reach the blastocyst stage can be implanted in the uterus of a **surrogate mother** for gestation. Most of the implanted embryos don't develop or survive to birth; of the 29 implanted embryos, only Dolly was born.

Similar problems have occurred with other animal species that have been cloned. The process is simply inefficient and generally the success rate ranges from 0.1% up to 5.4% (the 5.4% was reported from a SCNT using mouse adult skin cell nuclei). That means that, for every 1000 nuclear transfers into eggs, at best, 54 clones might be born, depending on the species and donor cell type. Why are the numbers so low? There are several steps that often don't work: if the egg with the new nucleus does not begin to divide or develop normally, if the embryo does not implant correctly into the uterus of the surrogate, or if the fetus dies before it has reached full term.

What about reproductive cloning in primates? Unlike other mammals, primates are far more challenging to clone using SCNT. In 2001, Advanced Cell Technology announced that it had performed SCNT with human cells. The embryos only developed to a 6-celled stage. SCNT has not worked in monkeys either. Upon further investigation of the early stage embryos produced with SCNT using monkey or human cells, it was found that the primate cells had major problems with the number of chromosomes in each cell. Some cells had normal chromosome numbers, but some had too many, too few, or none at all. Researchers now know that essential proteins in primate eggs are lost during the SCNT process. Without these proteins, chromosomes are not equally distributed to daughter cells during cell division. In June of 2007, advances in the SCNT procedure gave rise to healthier monkey embryos. What was new to this study was that the scientists performed the nuclear transfer using a light source to visualize the eggs; they believe this modification was less damaging than the earlier technique used with primate cells. Rather than implanting the embryos for live births, the embryos were dissociated and two ESC lines were established. Additionally, news in January 2008 from Stemagen, a company based in La Jolla, California, suggests that SCNT is looking more promising with humans, too. They announced that they had successfully cloned a human to the blastocyst stage using SCNT. These recent experiments demonstrate that some of the early technical barriers to reproductively cloning primates have been overcome.

Somatic cell nuclear transfer is not the only way to make clones. Scientists have shown in animals such as mice and monkeys that if the nucleus from an embryonic cell (rather than from a fully differentiated somatic cell) is injected into the enucleated egg (termed **embryonic cell nuclear transfer,** or **ECNT**) that the cloning efficiency is significantly greater. The reason for this is likely because embryonic cells are already capable of giving rise to many different cell types. Using cells from the same embryo for each nuclear transfer, this process can produce many identical clones of the embryo (Figure 14A). Note that, unlike SCNT, you cannot clone an adult this way. This process is more like a "multiples" birth, in which the offspring are clones of each other (such as identical triplets are identical to each other). So while SCNT has not been very successful with primates, in 1997, two monkeys were born via the ECNT process performed by Dr. Gerald P. Schatten and colleagues at the Oregon Regional Primate Research Center.

A second alternative to SCNT to clone an animal is the way in which Tetra, a female rhesus monkey, was cloned in 2000, also by Dr. Gerald P. Schatten and colleagues through a process known as **embryo splitting** (Figure 14B). The process involves normal fertilization

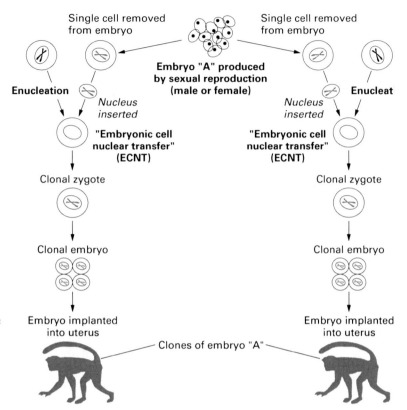

FIGURE 14A The process of embryonic cell nuclear transfer (ECNT) can make several identical clones if several cells from the same embryo are used. *Source:* Figure adapted from Thieman and Palladino, *Introduction to Biotechnology*, p. 261, Figure 11.20.

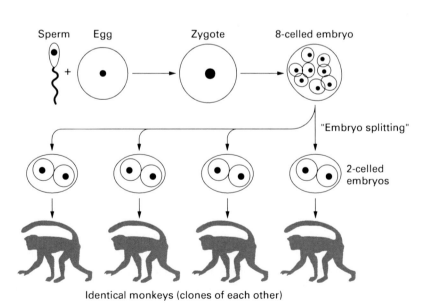

FIGURE 14B The process of embryo splitting produces (at best) four identical clones.

26 STEM CELLS AND CLONING

by a sperm and egg and allowing initial development. When the embryo reaches the 8-cell stage, the embryo is split into four 2-celled embryos that can give rise to four genetically identical clones. This technique can be thought of as a way to produce "artificial quadruplets." Embryo splitting is similar to ECNT in that an adult cannot be cloned by this technique but several identical clones could be made. Tetra was the only one of the four clones to survive until birth and is the first primate produced by this cloning technique. Scientists trying to clone monkeys suggest that having identical monkeys is useful for medical tests because they represent the closest animal model for human diseases and reduce the variability of using nonclones. For example, several animals with the identical immune systems would be useful in testing vaccines for HIV because it would rule out genetic diversity as a variable.

The Health of Reproductive Clones

We have discussed how very few clones survive and grow to adulthood. Are any of the clones that survive normal? Although there is not enough evidence to make a concrete conclusion about Dolly the sheep, many scientists believe that she aged prematurely. When scientists studied her telomeres ("caps" at the ends of her chromosomes) they found that they were slightly shorter than would be expected in a sheep of her age. Her telomeres suggested that her "genetic age" was greater than her true birth age, fitting with the reality that she was cloned from a 6-year-old sheep. And yet, shortened telomeres have not been consistently observed with other cloned animals. Other circumstantial evidence supports the idea of premature aging in Dolly—she developed early onset arthritis and died at the young age of 6 (sheep typically live 11 to 12 years) from a progressive lung disease. When we consider other cloned animals of various species, it is evident that making it to birth does not necessarily mean the animal is normal. Other problems seen in cloned animals are weak immune systems, tumor growth, and death at a young age. Additionally, reproductive clones and their placentas sometimes grow faster and larger than normal, termed **large offspring syndrome.** These large clones would need to be delivered by caesarian (C-) section. It is not known exactly why this overgrowth occurs, but it might have something to do with the way different genes are expressed during embryological development.

Normal development is an intricate dance of genes being expressed at the right time in the right cells. The difficulty in getting normal clones has a lot to do with gene reprogramming when the nucleus is transferred to the enucleated egg. In sexual reproduction, the egg and the sperm each have their genes programmed to start normal development. The way that genes are set to be expressed is called **gene imprinting.** Think of genes as a set of on/off switches; different switches will be set "on" or "off," depending on the particular cell type, tissues, or stage of development. When the nucleus of an adult somatic cell is transferred into the enucleated egg to create a clone, many of the switches are set differently than they would be for the start of embryonic development because by then the body cell has its programming set to do its job in a particular differentiated tissue (e.g., skin). The cytoplasm in the egg must **reprogram** the genes, resetting the switches so that normal development can occur. The failure of most reproductive cloning attempts is likely a result of not getting all the switches set correctly. In fact, one study examined the imprinting patterns for five specific genes in mouse embryos generated by SCNT. They found that only

4% of the cloned mice mimicked the blastocyst mode of expression for all five genes. These types of reprogramming errors also likely account for large offspring syndrome and other problems seen with many different species of clones. The reason that ECNT may be more efficient than SCNT is that embryonic cells likely need less reprogramming.

THERAPUETIC CLONING

Therapeutic cloning (research cloning) involves cloning to produce embryonic stem cells (ESCs) for medical therapies. The idea is to provide therapies for patients with diseases such as Parkinson's disease, diabetes, and so on, using ESCs. By performing SCNT with an enucleated egg, a cloned embryo of the patient (see Figure 13) is produced. Instead of implanting the embryo, ESCs would be derived from the blastocyst stage and grown *in vitro*. Theoretically, these cells would be a perfect tissue match, getting around the problem of transplant rejection. The ESCs would be stimulated in the laboratory to differentiate into healthy tissues. A person with Parkinson's disease might receive nerve cells to cure the disease, and an individual who has suffered a heart attack might receive heart muscle cells to repair his/her damaged heart. It should be noted that even if the cloned ESCs are never used in patients, the technique is still a powerful one for basic research. For example, the molecular mechanisms that cause many inherited diseases are unknown. Scientists could derive various ESC lines from individuals with genetic diseases and watch exactly how these cloned cells differentiate and function compared to normal cells. Pharmaceuticals could also be screened on these cloned cells to test their ability to slow or reverse the disease process.

The idea of therapeutic cloning is still somewhat theoretical and poses many problems. From a technical standpoint, it is still not possible because it would require SCNT to work successfully with human cells to consistently produce healthy cloned blastocysts for ESC retrieval. The first successful production of *some* such blastocysts was done with primates in June of 2007, and it required 278 eggs to obtain just two stem cell lines from cloned rhesus monkey embryos. News in January 2008 from Stemagen suggests that therapeutic cloning may be closer to reality in humans. They announced that they had successfully cloned a human to the blastocyst stage.

Many Eggs Are Needed for Cloning

Eggs are necessary to make clones. Not just any cell will do, since the egg's cytoplasm contains materials that activate the various genes to send the embryo through development properly. However, it takes more than one egg to make a clone. As we discussed previously, cloning is an inefficient process, even just to get the clone to the blastocyst stage where stem cells can be harvested. At best, the current success rate to clone a human embryo to the blastocyst stage from a pool of eggs is approximately 10%. This rate is based upon Stemagen's announcement regarding SCNT to produce cloned human blastocysts. With the current technology, many human eggs will be needed. Who will donate the eggs? To get numerous eggs (more than one per month), a woman must receive high doses of hormones to cause multiple eggs to mature, and surgery is then required to retrieve the eggs. The hormones can have serious side effects, and surgery poses risks to the egg donor, as well. In addition, if women are willing to donate eggs, what would the cost of this be? In 2003, a study was published that reported it would cost up to $200,000 per patient just to pay for the human eggs needed to derive one usable ESC line that could be used for therapeutic cloning. The high cost of developing a procedure that will undoubtedly remain costly for each patient has made some venture capitalists hesitant about investing in this field of biotechnology. Hence, it may not be possible or practical to get enough human eggs for cloning attempts.

One proposal is to use cow or rabbit eggs, instead of human eggs. These are readily available and can be harvested in greater numbers. Naturally, there is some concern about mixing species, even though the hybrid embryo will not be brought to birth. There is still some uncertainty if this would work for the patient receiving the resulting cells, since there would be a slight genetic contribution from the cow or rabbit egg.

A clone produced by nuclear transfer is *virtually* genetically identical to the donor genetic material. The "virtually" part arises because there is some contribution from the egg. The genetic contribution comes not from the egg's chromosomes, which are removed during enucleation, but from the mitochondria, the little energy-generating factories in the cytoplasm of each cell. These are still present in the egg when the donor nucleus is transferred in. Even though mitochondria do not contain much DNA, there are lots of them to make energy for the cell. As much as 1% of the DNA in the clone may be derived from the egg in this way. Some of the proteins made from these genes do wind up on the surface of the cell, where there is the possibility that they could trigger an immune response. While the chance of this may be low using human eggs, it could cause a significant problem if cow or rabbit eggs are used.

Before we move away from this topic of therapeutic cloning, it is worthy to note a news story that made international headlines in 2005. A South Korean researcher, Hwang Woo Suk, published a report in a world-renowned science journal in which he claimed that he had derived 11 human ESC lines from cloned human embryos. Many scientists were excited by the news that, not only was he able to perform SCNT on human eggs, but he had derived ESC lines from them. At the time, it appeared that therapeutic cloning was a reality and it was just a matter of time before it proved useful as a clinical therapy. Not long after his publication appeared, however, he was exposed for falsifying the data presented. After much investigation into his studies, it became clear that he had not successfully created cloned human ESCs nor had he even been successful at SCNT to produce blastocysts. This publicity caused much skepticism about stem cell research among the public. When Stemagen announced that they had successfully cloned a human to the blastocyst stage using SCNT, they validated their results with accompanying evidence that it had worked. Yet because of the previous Hwang scandal, many scientists remain skeptical about the news. It remains to be seen if Stemagen can produce blastocysts that are healthy enough to produce cloned human ESCs.

Bioethics of Stem Cells and Cloning

BIOETHICS OF STEM CELL RESEARCH

The real root of the debates regarding stem cells is the question of moral status of the human embryo. Scientifically, the embryo is a human being, just starting out on its developmental journey, but the science has no answers for the deeper questions regarding how we view this tiny entity. Instead these are moral, philosophical, and theological questions. Does it have a soul? Is it conscious? Is it a person or a piece of property? When does human life begin? On the one hand, a human blastocyst is a mere dot (it fits into Roosevelt's eye on the face of a U.S. dime). It does not possess a beating heart or brain waves; it is without arms and legs. Of course those things develop later; but, at the very early stages, the embryo lacks these things that we usually associate with our idea of a person. There are various views but no consensus of the status of the early human embryo. Some say it is simply a clump of cells, just like a chunk of skin. Others believe that it is a form of human life, deserving of profound respect yet only a *potential* person. Still others maintain that an embryo has the same moral value as any other member of the human species. How individuals answer these questions may be based on religious views. Interestingly, defining when life begins is different for various faiths, and so some religions are more tolerant than others of aspects of this research. For example, the Jewish faith believes that humanhood begins at a later stage of development than the blastocyst, and it does not grant legal status to an embryo. As a result, groups such as Hadassah, the Women's Zionist Organization of America, and the Union of Orthodox Jewish Congregations of America are outspoken supporters of ESC research for the purpose of curing disease. In contrast, the Catholic Church believes that at the time of conception (either *in vitro* or *in vivo*) the embryo is a human being with full rights. The Catholic Church opposes the direct destruction of blastocysts for any purpose, including research. Faith is just one of the many factors that might shape a person's opinion on these moral questions. The news, education, personal nonreligious beliefs, and personal experiences are other major factors shaping people's views about stem cells.

The concept of using human embryos for research and potential stem cell cures started with the so-called "excess" embryos, left in the freezers at fertility clinics. For *in vitro* fertilization (IVF), a woman is given high dose of hormones to mature many eggs at once—ten, twenty, even thirty. These eggs are then surgically harvested and fertilized in a laboratory dish. The purpose of IVF is to help a couple conceive a baby. After fertilization and growth in a dish for a few days, anywhere from one to six of the embryos are implanted in the woman's uterus, and the rest of the embryos are frozen. If the couple is not successful with the first round, the frozen embryos may be thawed and used for a second attempt. Or, after the initial birth, the couple may want to use the frozen embryos to produce more children. Nevertheless, there are inevitably some embryos that are not used and are left in the freezer. They can survive for extremely long periods of time.

Inevitably, some of the frozen embryos are discarded. Fertility clinics obtain a signed consent form from the couple, and one option allows the clinic to discard embryos after several years if they are not used. However, most are not discarded. What to do with all the frozen embryos? One option for frozen embryos is relatively new—embryo adoption. The Snowflakes Embryo Adoption program sets up adoptions for embryos whose genetic parents have achieved their family. However, it is estimated that there are 400,000 excess embryos in freezers around the country, so embryo adoption may not make a big impact on this number. Newer techniques often don't create as many embryos in the past. For example, it is possible to only create as many embryos at one time as will actually be implanted or to freeze just the eggs, which can be thawed, fertilized, and the embryos implanted as needed.

Nonetheless, there *is* an excess of embryos from IVF. Another option for parents: their embryos may be donated for research. This was the bone of contention in the stem cell debate. Should human embryos be used for research, given their original purpose was to be implanted for a birth? Why not use them for research if they would be discarded (with the parent's permission) anyway?

When considered in the context of using human embryos for their ESCs, the necessary destruction of the embryo is a contentious one. Should human embryos be destroyed if, from that destruction, it might be possible to treat many patients suffering from disease? Are embryonic stem cells as good for potential treatments as claimed, or are there still viable alternatives that are just as good? Or is the research still necessary so that science can explore all possible avenues for medical breakthroughs? Do the ends justify the means?

ALTERNATIVES TO DESTROYING EMBRYOS FOR STEM CELL RETRIEVAL

"...although adult stem cells may not provoke much political rancour today, they have become more scientifically controversial than their embryonic counterparts" –Christine Soars (SCIAM, July 2005).

The quote above states the dilemma clearly. The view of many scientists at this point is that adult stem cells alone just won't do. For example, finding and characterizing these cells has been difficult and not everyone is in agreement about their abilities to differentiate and even transdifferentiate. For this reason, the push goes on for work with ESCs. However, some scientists are trying to find ways around destroying an embryo; they are attempting to find alternative sources for ESCs in order to skirt ethical issues. How scientists have approached this dilemma demonstrates just how creative they can be as a group—when presented with a problem, scientists will attempt to find many ways to solve it. One way around the ethical dilemma of destroying embryos for their unique pluripotent cells would be to find another cell type that behaves like or can be coaxed into behaving just like embryonic stem cells. Let's examine some of these alternatives.

Amniotic fluid–derived stem (AFS) cells have recently been well characterized and demonstrate similarities to ESCs. Amniotic fluid–derived stem cells show pluripotentiality and, similar to ESCs, have been differentiated into cells representing all three germ layers, such as muscle, brain, and liver cells. A unique aspect of AFS cells is that they grow fast like ESCs but do not show signs of aging or developing into tumors, even after two years of *in vitro* growth. While they may not provide an alternative for all types of research, they may prove to be a suitable choice for many uses. These cells can be easily collected from amniotic fluid, as early as 10 weeks after conception, and from the placenta, which is often discarded at birth. Similar to umbilical cord blood, these cells could be saved at birth for an individual's potential use later in life. Of course, privately banking stem cells is costly on a per individual basis. However, a bill has passed in the Senate and awaits a House committee that would provide funds to establish national banks to maintain amniotic fluid and placental stem cells. One-hundred thousand samples has been suggested as the approximate number of samples needed to account for the genetic diversity needed to provide enough immunologically compatible tissues for everyone in the United States. However, a bill has been introduced in the U.S. Congress that would provide funds to establish national banks to maintain 100,000 amniotic fluid and placental stem cell units. This many samples has been suggested as the approximate number of samples needed to account for the genetic diversity needed to provide enough immunologically compatible tissues for everyone in the United States.

What if other cells could be coaxed into behaving like ESCs? There is evidence that this is possible scientifically and support for it ethically. A stem cell may turn out to be not an entity so much as a "state"—one that any cell could enter under the right conditions. Along these lines, some groups are trying to determine what genes must be expressed to make a cell behave like a pluripotent ESC. In 2006, scientists from Japan, Kazutoshi Takahashi and Shinya Yamanaka, first described **induced pluripotent stem (iPS) cells**—differentiated cells that are reprogrammed to become pluripotent *in vitro*. Using a virus to introduce just four genes (*c-Myc, Oct3/4, Sox2, Klf4*) into mouse skin cells, they showed that their iPS cells were similar to ESCs; they could differentiate into all three germ layers *in vivo* and *in vitro* (Figure 15). In late 2007, the Japanese scientists and an independent group at the University of Wisconsin–Madison announced that they had used the same method to produce human iPS cells from differentiated skin cells. Significant technical challenges may still need to be overcome to apply this technology, but this type of experiment demonstrates that it may one day be possible to avoid destroying embryos for their

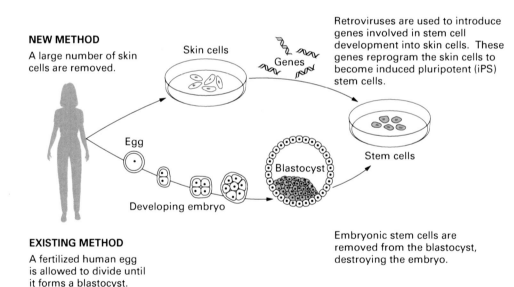

FIGURE 15 Reprogramming human skin cells. Researchers have developed a technique for creating stem cells (induced pluripotent stem [iPS] cells) without the controversial use of human eggs or embryos. If the method can be perfected, it could quell the ethical debate troubling this field. *Source:* Figure adapted from New York Times Website http://www.nytimes.com/2007/11/21/science/21stem.html?_r=1&sq=embryonic+stem+cells&st=nyt&oref=slogin.

unique pluripotent cells. Even outspoken opponents to ESC work find this new technique ethically acceptable. For example, Richard Doerflinger, the U.S. Conference of Catholic Bishops' spokesman on stem cell issues states it *"raises no serious moral problem because it creates embryonic-like stem cells without creating, harming, or destroying human lives at any stage."* So while scientists creatively develop techniques to avoid ethical issues, they additionally gain a better understanding for the cellular "state" of pluripotency.

Recently, a promising innovative treatment using iPS cells to correct sickle cell anemia in mice was reported. iPS cells were produced from skin cells of mice that express a mutated version of the human sickle cell hemoglobin gene and display sickle cell disease. The iPS were genetically engineered to correct the hemoglobin gene mutation. Blood stem cells were then made from the corrected iPS cells and transferred into donor sickle cell mice which produced functional red blood cells that corrected the disease condition. This is an incredibly exciting result combining aspects of both stem cell technologies and gene therapy to treat a genetic disorder.

Another method designed to avoid the ethical dilemma of destroying an embryo for ESCs is to "pluck" one cell, termed a **blastomere,** from an 8-celled embryo. The single cell would be developed into an ESC line while the now 7-celled embryo would go on to develop into a healthy baby. In 2006, scientists at Advanced Cell Technology demonstrated that it is possible to derive ESCs from this one cell, although these "proof of principle" experiments demonstrated how technically challenging this could be. They have since improved the process and established five more cell lines. But how do we know the 7-celled embryo can go on to produce a healthy baby? A process called preimplantation genetic diagnosis provides evidence. **Preimplantation genetic diagnosis (PGD)** is a technique sometimes used to determine if an embryo generated by *in vitro* fertilization has a genetic defect. During PGD, one of the eight cells is used for genetic testing (which destroys the cell), but the 7-celled embryo successfully develops into a viable human. This has been successfully done for many years. The significance of the experiments by Advanced Cell Technology is to demonstrate that ESCs can be derived from an embryo without destroying it. Nonetheless, opponents to this method for deriving ESCs state that it is still unethical because the procedure still poses a needless risk of death to the embryo.

Would the ethical dilemma be avoided if ESCs were removed from embryos that are not inherently capable of developing past a particular stage of development? A method such as this was introduced in 2005 at the Massachusetts Institute of Technology. In these studies, they produced a mouse that has a gene defect such that the embryo cannot attach to the uterus wall. Without attachment, they are not considered to be viable. These same ideas could then be applied to defective human embryos. The human embryos would develop normally for the first few days but would not have the ability to continue development without attachment. Would removing cells from these types of embryos be considered ethically acceptable? It appears there is little support for this method as an ethical alternative. As stated by a representative from the U.S. Conference of Bishops, *"Creating these human lives just to destroy them is wrong,"* he said. *"Engineering them so they will self-destruct after a certain point in development is wrong."*

Another source of embryonic stem cells that may prove to be ethically acceptable was reported in June 2007 by scientists at Lifeline Cell Technology of Walkersville, Maryland, and from Moscow, Russia. In these experiments, unfertilized human eggs were stimulated to begin dividing, a process known as **parthenogenesis,** and were grown to the blastocyst stage (scientists don't believe these embryos have the potential to develop long enough to produce babies). Embryonic stem cell lines were then derived from these parthenogenic blastocysts. You are probably wondering how an embryo can be made if there is no fertilization. It turns out that normally before fertilization, the egg actually still has a full number of chromosomes—half will be used to merge with the half brought in by the sperm, and, upon fertilization, the other half of the egg's chromosomes will be shot out of the egg in what is called a **polar body.** For parthenogenesis to occur, the egg is induced by chemicals to keep both sets of chromosomes, resulting in a full number of chromosomes inside the egg. Some animals can reproduce this way in nature, such as salamanders, some reptiles, birds such as chickens, and some insects. When used in humans, this technique can produce ESCs that are not true clones of but are genetically matched for the woman who donates the egg. With drugs to suppress the immune system, patients other than the female donor might be able to use these cells as well. Technical issues still need to be resolved, such as determining if these cells are genetically flawed and if they can be safely used for treatments. Since some people see these as only "activated" eggs, they are not morally conflicted. Others may see these embryos as defective but human nonetheless.

Another proposal that has been discussed for avoiding the destruction of embryos to obtain ESCs is to remove living stem cells from surplus IVF embryos that are not healthy enough to develop further (i.e., are developmentally arrested). Two scientists at Columbia University, Donald Landry and Howard Zucker, propose that some embryos with severe genetic defects that are arrested in development have individual cells that are healthy. These cells could be harvested from the embryos analogously to a situation in which organs are removed for transplantation from brain-dead individuals. Current technical methods still need to be developed to distinguish embryos that are healthy versus those arrested in development versus those that are dead.

BIOETHICS OF CLONING

Human cloning raises even more questions about what it means to be human. Cloning involves the specific creation of human embryos with certain ends in mind. For reproductive cloning, there are worries beyond the safety factors. For a couple who create a cloned child, a clone of the wife will not be the genetic daughter but instead will be the sister, a late-born twin, and will not be related to the husband. How will this change the kinship and family relationships? What will be the expectations put on a clone once born? And what if—and this is possible—a previously existing person, now deceased, is cloned? The genetic makeup of a clone will already be known, already dictated because the process of reproductive cloning reproduces a previously existing individual. Will the clone be expected to live up to that genetic legacy? Will there be heightened expectations by the parents and others based on what was achieved by the donor of the genetic material that has made the clone?

However, keep in mind that even though the genetic makeup of the clone is predetermined, there are many other factors that go into our overall composition. Our genes determine many of our physical characteristics and even predispose us to various diseases or behaviors, but we are also products of our own environment and experience. A good example is the first cloned cat, "cc" (short for "carbon copy"). Even though she looks very similar to the cat who donated her genetic material, her coat pattern is slightly different (Figure 16). This is because even the environment in the womb can affect development (in this case, coat pattern). Even identical twins have different fingerprints. Of course after we are born, there are many experiences that make us who we are. Those experiences can't be duplicated, so the clone will grow up differently than the one who was cloned and may behave quite differently. A clone of Einstein might become an artist instead of a scientist. We are so much more than just our genes!

Creating cloned embryos for experimental or medical use raises the questions previously posed regarding moral status. Should humans (realized or potential) be created and destroyed for the potential benefit of others? Again, there is a range of viewpoints regarding the status of the embryo. If the embryo at this early stage does not have as high a value, or any value, compared to other human life, then **utilitarian logic** (the idea that something is only good if it is useful and that actions should promote the greatest good for the greatest number of people) would dictate that it should by all means be used. Others argue that it should be protected, not because of its inherent value but because creation of human embryos for such purposes can lead to human commercialization, making any human life a commodity to be bought, sold, and used, thereby cheapening the value of life. Still others would say that we should not create human embryos for purposes other than reproduction and not in a manner that manufactures human life, so-called "designer embryos." The question still goes back to what it means to be human and what value is placed on human life.

Have you considered your own thoughts on cloning? Do your thoughts differ when considering humans versus animals? According to Gallup's 2007 Values and Beliefs Survey, conducted in May of 2007, when asked about reproductively cloning animals,

FIGURE 16 "CC" the cloned cat (*left*) and her genetic donor, Rainbow (*right*). Although they have the same DNA, they have slightly different coat patterns.
Source: Pat Sullivan/APImages (left), Alpha/ZUMAPRESS/Newscom (right).

36% thought this was morally acceptable. Yet, the view is different when the life is a human one. Most scientists and the general public oppose attempts to clone a human to produce a child. Only 11% of the American public believes reproductive cloning of humans is morally acceptable. The National Academies of Sciences has recommended that cloning humans should not be done at present, simply because it is "unsafe."

There is far more support for therapeutic cloning than reproductive cloning in the American public. Another poll asked by the Coalition for the Advancement of Medical Research (CAMR) found that, as people learned more about what therapeutic cloning was, they became more accepting. For example, when a group was asked about their opinion of cloning to develop stem cells, 60% strongly or somewhat favored the research. However, after reading a more detailed description of what therapeutic cloning research involves, 72% of the respondents favored the research.

Stem Cell and Cloning Policies

In such contentious and divisive issues, with prestige and money at stake, politics and policymaking are in play as well. The issues of stem cells and cloning have generated a great deal of debate, Congressional hearings, speeches, and rhetoric. The ongoing debate is helping everyone learn the issues and make their voices heard.

THE DEBATE IN THE UNITED STATES

In the United States, the main focus has been on federal funding. In 1996 under the Clinton administration, a legislative ban stated that federally appropriated funds could not be used for the creation of a human embryo for research purposes. The ban clearly defined "human embryo." However, once human ESC lines had been well established two years later in 1998, the legal definition of ESCs had to be considered; were these cells defined as "embryos"? In 1999, Harriet S. Rabb, then general counsel of the Department of Health and Human Services, provided her legal opinion to the National Institutes of Health (NIH). She stated that the ban did not apply to *isolated* pluripotent stem cells because they were not capable of developing into a human, even if placed in a uterus. Guidelines were established by the NIH that stated that pluripotent stem cells (ESCs) could be used in NIH-funded studies, provided that some other (nonfederal) entity paid for removal of the cells from human embryos. The NIH initiated the applications process, but ultimately funding was not granted to the applications because of the new president's policy. On August 9, 2001, President George W. Bush announced that only human ESC lines that had already been established *before* this date would be eligible for federally-funded research. At this time, it was believed that 78 lines met the qualifications set by President Bush. Subsequently, it became clear that many of these lines were either duplicates, failed to grow, or were held by labs that either withdrew them or refused to send them to U.S. researchers. As of March 2007, the NIH was reporting that 21 lines were available to researchers performing research with federal funds.

The limited number of cell lines has frustrated scientists who perform human ESC research for several reasons. Why? First, the limited number of lines means there is limited genetic diversity. To learn how "normal" ESCs behave would require many more cell lines, as would predicting how cells from various ethnic groups would respond to the same drug treatment. Second, cell lines can acquire genetic mutations over time, and this can render them useless for various applications. As these 21 lines age, this genetic instability becomes an increasing concern to scientists working with them. Third, the original cell lines were isolated and cocultured with mouse "feeder" cells and bovine serum. Contamination with nonhuman products makes these lines risky for future use in human therapies because they may incite serious immune system responses. (Since 2001, technology has significantly improved such that it is now possible to derive new ESC lines in defined conditions that completely lack animal products.) Last, not only do scientists want to study normal cells from "normal" embryos, but they want to study cells that contain genes for specific diseases. Being able to observe cells as they develop abnormalities may allow researchers to see exactly where development goes wrong and provide insight in correcting it. For all the reasons described, many scientists would like to see the ban lifted so that new ESC lines could be derived from surplus IVF embryos.

As of 2008, researchers in the United States using federal funds were still subject to the ban set by President Bush in 2001. However, several bills had been in Congress in attempts to overturn the ban. In May 2005, the House of Representatives passed H.R. 810, the **Stem Cell Research Enhancement Act,** which would expand federal funding to lines created *after* August 9, 2001, from excess embryos stored in fertility clinics. In July of 2006, the Senate also passed the bill, but it immediately fell to a presidential veto, President Bush's first veto in six years! (Neither the Senate nor the House could gather enough votes for the 2/3 majority required to override the veto). The Stem Cell Research Enhancement Act (renamed H.R.3) was reintroduced after mid-term Congressional elections. In late spring of 2007, both the Senate and House of Representatives approved the bill, once again. Despite even more support than for H.R. 810 there was still not enough support to override President George W. Bush's second veto of his presidency. This issue will certainly be raised again for the president-elect of 2008.

The U.S. public opinion reflects the votes in Congress. While some groups may loudly protest the use of medical research using stem cells obtained from human embryos, as a whole, it seems that Americans have grown more accepting. National polls attempt to take a "snapshot" view of current opinion across the country. According to one group, The Gallup Organization, which followed the public's views from 2002 to 2007, the percentage of Americans considering the use of ESCs obtained from embryos morally acceptable has gradually increased from 52% to 64% (Figure 17).

Although the development of stem cell lines with federal funds remains stalled at the time this text was written, it is legal—in most states—to produce new ESC lines, provided that the funding is from state or private sources. Many people believe that without the NIH, the largest supporter of biomedical science in the United States, the field is hampered and American scientists are at a disadvantage. Although private companies fund this research, too, these companies may closely protect their results for proprietary reasons to gain exclusive commercialization rights (patents) to the cell lines and technologies they have invented.

Patents are another controversial area within the field of stem cell research because some people argue that patents impede the ability of some research to move forward. Three patents specifically have been central in the news. These three key patents are held by the Wisconsin Alumni Research Foundation (WARF), which manages the intellectual property

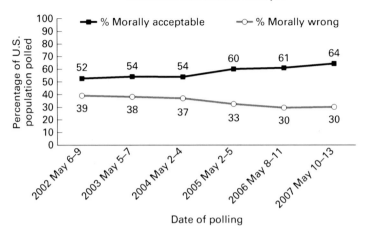

FIGURE 17 Since 2002, Americans have grown more accepting of medical research using stem cells obtained from human embryos. The percentage considering this morally acceptable has gradually increased from 52% to 64%. *Source:* Gallup Poll, http://www.galluppoll.com/content/default.aspx?ci=27757.

of University of Wisconsin–Madison. (Remember, it was James Thomson at Wisconsin who first derived human ESCs in 1998.) The patents protect Wisconsin's ESC lines and the methods used to grow them. The result is that WARF has charged biotechnology firms license fees up to $400,000 for permission to use the ESCs and/or methods. Academic labs even have to pay a small fee. However, the patents were challenged in July 2006 by the U.S. Foundation for Taxpayer and Consumer Rights (FTCR) and the Public Patent Foundation (PUBPAT) because much of the money used to support academic and institutional research comes from taxpayers. They state that these patents, "purport to cover stem cells that are looting taxpayer funds and forcing research overseas." As of 2008, only one of the three patents was upheld by the U.S. Patent and Trademark Office (PTO). The two others were revoked but are under appeals.

Despite the lack of NIH funds to public institutions, private donations to universities and institutions and initiatives in various states continue to push the field forward in the United States. California leads other states with the most dollars committed by taxpayers with a measure passed in November 2004. **Proposition 71, the California Stem Cell Research and Cures Act,** provides $3 billion for research and facilities and created the California Institute for Regenerative Medicine to regulate stem cell research. Other states that have committed to funding embryonic stem cell research are: New Jersey, Connecticut, Illinois, Maryland, Massachusetts, Indiana, and Wisconsin. Interestingly, while some states are encouraging stem cell research, others (such as South Dakota) prohibit research specifically on cells or tissues derived from an embryo outside a woman's body.

Until now, we have only discussed the politics of ESC research, but what about cloning? You might be surprised to find out that there are no federal laws against reproductive or therapeutic (research) cloning. In 2003, the House of Representatives passed a bill (Human Cloning Prohibition Act, H.R. 534) to ban both reproductive and therapeutic cloning, which President Bush indicated he would sign. The Senate tabled the legislation, and a vote was never taken. The Human Cloning Prohibition Act of 2007 (H.R. 2560) was introduced in the House by Representative Diana DeGette (Democrat–Colorado) and was voted down. The Act would have made it unlawful for "any person, including a governmental entity, to perform or attempt to perform human cloning." A Representative opposed to the bill (Joseph Pitts, Republican–Pennsylvania) stated that the bill, *"allows for unlimited cloning of human embryos but prevents women and doctors from trying to implant one of these embryos to initiate a pregnancy. In practice, this means that embryos will be cloned, used for experimentation, harvesting, research, then assigned a death sentence. So cloned embryos would be required by law to die. Not only does this bill allow the practice of cloning to move forward, it also mandates the killing of those human embryos."*

There is discussion in the Senate as well. In 2007, Sam Brownback (Republican, Kansas) reintroduced legislation in Congress that would make all human cloning, including therapeutic cloning, illegal in the United States. The legislation would also ban importation of any medical products created using the technology in other countries. Punishment would be up to 10 years in prison and a $1 million fine. An alternative act, introduced by U.S. Senators Orrin G. Hatch (Republican–Utah) would make reproductive cloning illegal, too, but permits SCNT for therapeutic cloning. Neither Senate bills had been voted on by mid-2008.

In the absence of federal legislation, various states have made laws pertaining to human cloning. Although some states have banned both reproductive and therapeutic cloning, others have banned only reproductive cloning. For example, New Jersey and California have both passed laws that specifically permit SCNT for the purpose of making ESC lines. So, as long as scientists are not using federal money, it is legal to derive ESCs from cloned embryos in these states. Almost all states are currently debating stem cell and cloning issues and have bills under consideration. The individual states may be the ones that move faster in terms of the legislation, which may affect the debate at the federal level.

WHAT ARE OTHER COUNTRIES DOING?

Not surprisingly, stem cells and cloning are issues for debate not just in the United States but around the world. Many countries are establishing stem cell research centers, including the United Kingdom, France, Germany, Switzerland, Belgium, Sweden, China, Taiwan, South Korea, Japan, India, Australia, Turkey, and Israel. As you might expect, there is no consensus about policies regarding existing ESC lines or establishing new cell lines. Some countries don't allow any work with human ESCs, some only allow work with human ESCs if they are imported from another country, and others have vague or very relaxed laws about the sources of human ESCs. Much is at stake in terms of possibilities for scientific advances, economic and biotechnology development, and national prestige. In September of 2003, the InterAcademy Panel on International Affairs (IAP) published a statement that called "on all countries worldwide to ban reproductive cloning of human beings" yet requested that cloning for research and therapeutic purposes be excluded from the ban. More than 60 science academies from across the world signed the statement. Although there is international agreement that reproductive cloning should be banned, therapeutic cloning continues to be controversial.

In early 2005, the United Nations (UN) General Assembly passed a nonbinding declaration banning all human cloning, including therapeutic cloning. The vote was 84 to 34 (with 37 abstentions), reflecting the partition over the issue worldwide. The United States was an outspoken supporter of the UN ban. Some countries already have laws reflecting this sentiment. For example, our next-door neighbor, Canada, passed legislation in 2004 that specifically prohibited both types of human cloning. Nonetheless, there are many stem cell advocates. The United Kingdom (UK) legalized SCNT in 2001 and, despite the UN's declaration, continues to support this type of research. Several other countries, such as South Korea, Australia, Belgium, Israel, India, and Singapore, also allow SCNT, and scientists from these countries are attempting to establish ESC lines from cloned embryos.

Permitting therapeutic cloning necessitates further policy, as outlined by the first international guidelines issued by the International Society for Stem Cell Research (ISSCR) in February of 2007. Two key issues are highlighted in these nonbinding guidelines: 1) payment for women donating eggs, and 2) control on animal-human chimeras. As you now know, the inefficiency of therapeutic cloning requires a large source of human eggs for nuclear transfer. The UK leads on this issue with a law prohibiting the use of cash payments to induce women to donate eggs for research. One research institute in the UK now offers an "egg-sharing" scheme: in return for donating spare eggs, women will be offered contributions towards expensive IVF treatment. Another area of policy that "cloning permissive" countries must face involves controls on projects involving animal-human chimeras. In lieu of human eggs for nuclear transfer, animal eggs might be a more readily abundant source. The resulting embryos would then be almost entirely human, with only animal DNA outside of the nucleus. Policymakers in the UK are currently debating these issues, as a cow-human hybrid proposal was submitted in late 2006 for the purpose of creating ESC cell lines.

The Future

If you follow the media related to stem cells, it appears at times that there is a war in which adult-derived stem cells are against embryonic stem cells. Which will win? Are adult or embryonic stem cells the future of this field? The simple answer is that it depends on the scientific application. For example, if the purpose of using stem cells is to regenerate tissues in a patient, the "best" cells might be those least likely to cause immune complications, that is, the patient's own adult-derived stem cells. However, if the intent of use for stem cells is to better understand the basic science of pluripotency, the "best" cells might be embryonic stem cells. As we view this exciting time in stem cell biology, we should consider each type of stem cell for its own unique characteristics. And remember, while it has been traditional to think about stem cells as adult or embryonic, the cells that we have discussed will not fit so neatly into two categories. In actuality, there is a continuum of stem cells (e.g., embryonic, fetal, amniotic, adult tissue) with different characteristics that are best suited for different applications.

What makes the future of stem cell science (and cloning) so unique compared to other biological advances is that the ethical and political issues are so closely tied to each experiment. It is precisely these political and ethical "limitations" that have pushed scientists to explore acceptable "alternatives" to embryonic stem cells. We have seen demonstrations of inventiveness and creativity in the scientists seeking alternative sources of pluripotent cells. For example, scientists are now learning how to induce pluripotency in adult cells. In the future, might a patient be able to give a few skin cells, have them induced to a pluripotent state, and then have these cells differentiated into any tissue that needs repairing? If so, it would avoid the destruction of embryos as a source of ESCs and would obliterate the need for the inefficient SCNT process for the purpose of "customizing" cells. The alternatives (already discovered or soon to be) may hold the key to the future of regenerative medicine; they may provide an abundant source of pluripotent cells that are easily obtained, cost-effective, and morally acceptable to all.

Because the moral and political issues may guide the science, it is difficult to predict where this nascent field of stem cell biology is going. We are left with a lot of questions at this time. For example, will the United States be a leader in the field of stem cell science, or will regulations hinder advances? Will ESCs even be necessary or practical in regenerative medicine if a pluripotent adult cell can be identified or induced? Will therapeutic cloning change the face of medicine, or will some other technology serve the same purpose? Will humans eventually be reproductively cloned? In the current climate, it seems important to many, but certainly not all, scientists that there be exploration of all types of stem cells for various research and therapeutic applications. (Even if U.S. policy doesn't permit this, other countries will support it.) Nevertheless, the science will eventually align with what works best technically, morally, and financially. It is safe to say that stem cells will change regenerative medicine immensely in the next few decades. However, the nature of these stem cells and the methods that will be used are not obvious yet. I anxiously await the new advances and hope you will continue to follow this ever-changing field that is so carefully scrutinized by all.

                        Kelly A. Hogan
                        Department of Biology
                        University of North Carolina at Chapel Hill
                        Chapel Hill, NC 27599-3280
                        Email: Kelly_Hogan@unc.edu

Resources

WEBSITES:

(Note: Because the information of these resources is relatively new and changing rapidly, websites provide some of the best up-to-the-minute information regarding background as well as current knowledge. The following are some of the best across a wide spectrum of view points.)

TUTORIALS/EDUCATIONAL INFORMATION

National Institutes of Health Stem Cell Information (http://stemcells.nih.gov/index. asp) Basic information on stem cells, research guidelines, approved stem cell lines, and other information. Two useful reports: Stem Cell Basics (http://stemcells.nih.gov/info/basics/) and Regenerative Medicine 2006 (http://stemcells.nih.gov/info/scireport/2006report.htm).

The National Academies (http://dels.nas.edu/bls/stemcells/booklet.shtml) Understanding Stem Cells: An Overview of the Science and Issues

National Center for Case Study Teaching in Science (http://ublib.buffalo.edu/libraries/projects/cases/case.html) The Center provides a site to search for educational case studies related to various topics in science including many related to regenerative medicine, cloning, stem cells, and genetics.

ANIMATIONS/VIDEOS

University of Michigan's Interactive Tutorial: Stem Cells Explained (http://www.lifesciences.umich.edu/research/featured/tutorial.html)

Howard Hughes Medical Institute: Biointeractive (http://www.hhmi.org/biointeractive/stemcells) Many videos, lectures, animations related to stem cells and cloning.

University of Utah Genetics Learning Science Center (http://learn.genetics.utah.edu). Lessons, quizzes, and animations related to stem cells and cloning.

Riken Center for Developmental Biology (http://www.cdb.riken.jp/jp/stemcells/) A Japanese Research Foundation with excellent animations related to basic characteristics of stem cells.

Sumanas, Inc. (www.sumanasinc.com/webcontent/anisamples/nonmajorsbiology/stemcells.html) A basic animation related to isolating embryonic stem cells.

PBS Nova: Science Now (http://www.pbs.org/wgbh/nova/sciencenow/3209/04.html) Video clip explaining therapeutic cloning, and various interviews, photographs, issues are highlighted at the site.

NEWS

Yahoo News Full Coverage: Stem Cell Research (http://news.yahoo.com/fc/Science/Stem_Cell_Research) Provides links to news, editorials relate to stem cell research.

Stem Cell Research News (http://www.stemcellresearchnews.com/) Independent reporting organization providing current objective headline news on all facets of stem cell research.

***New Scientist* Special Report on Stem Cells** (http://www.newscientist.com/channel/sex/stem-cells) Website for *New Scientist* magazine with current news and information related to stem cells and cloning.

***Scientific American's* Science News** (http://www.sciam.com/news_directory.cfm) Updated daily with science news stories. Search for news related to stem cells and cloning.

National Public Radio (http://www.npr.org/) Search for short news stories and audio interviews related to stem cells and cloning.

POLICY

National Institute of Health (http://stemcells.nih.gov/policy/guidelines.asp) See a listing of current and older official documents related to U.S. policy relating to stem cells.

StemGen (http://www.stemgen.org/) A website providing a world map and searchable database about laws and policies relating to stem cell research.

National Conference of State Legislatures: State Embryonic and Fetal Research Laws
(http://www.ncsl.org/programs/health/genetics/embfet.htm) This website is updated often and provides text and a table comparing the laws relating to stem cells and cloning for various states.

INTEREST GROUPS

(*Note:* These groups have an agenda and the information provided by them is not always objective!)

The Coalition for the Advancement of Medical Research (http://www.camradvocacy.org/) Consists of patient organizations, universities, scientific societies, and foundations advocating the advancement of breakthrough research and technologies in regenerative medicine.

The Royal Society (http://www.royalsoc.ac.uk/landing.asp?id=1202) The Royal Society is the national academy of science of the UK and the Commonwealth. They endorse research for adult-derived stem cells and therapeutic cloning and support a ban of reproductive cloning.

Americans to Ban Cloning (http://cloninginformation.org) A group of concerned Americans and U.S.-based organizations that promote a global ban on human cloning.

Do No Harm: The Coalition of Americans for Research Ethics (http://www.stemcellresearch.org/) A national coalition of researchers, health care professionals, bioethicists, legal professionals, and others dedicated to the promotion of scientific research and health care that does no harm to human life.

Juvenile Diabetes Research Foundation International (http://www.jdrf.org/index.cfm?page_id=103932) A charitable funder and advocate for type I diabetes research worldwide. The foundation supports and expansion of embryonic stem cell research.

Genetics Policy Institute (http://www.genpol.org/) A nonprofit organization dedicated to establishing a positive legal framework to advance stem cell research

Catholic Church (http://www.americancatholic.org/ and http://www.catholicnewsagency.com/) The Catholic Church is against embryonic stem cell research because it involves the destruction of human embryos. Search for articles on topics relating to stem cells and cloning.

BIOETHICS

Bioethics.net. The American Journal of Bioethics Online (http://bioethics.net/) Contains lots of articles and basic discussion on bioethics issues relating to stem cells and cloning.

The Council for Responsible Genetics (http://www.gene-watch.org) A nonprofit, nongovernmental organization that fosters public debate about the social, ethical, and environmental implications of genetic technologies. Search for articles related to stem cells and cloning in their bimonthly magazine, *GeneWatch*.

Bioethics.com (http://www.bioethicsnews.com/) A global information source on bioethics news and issues. You can vote on polls and see current results too.

The President's Council on Bioethics (http://www.bioethics.gov/) Established in 2001, the Council contains members appointed by the President for the purpose of advising on bioethical issues that may emerge as a consequence of advances in biomedical science and technology.

ARTICLES:

Cookson, C., et al. (July 2005). The future of stem cells. *Scientific American and Financial Times* Special Report. July 2005 (pages A1–A35).

Gilbert, D. (2004). The future of human embryonic stem cell research: Addressing ethical conflict with responsible scientific research. *Medical Science Monitor,* 10, RA99–103.

Hanna, J., Wernig, M., Markoulaki, S., et al. (2007). Treatment of sickle cell anemia mouse model with iPS cells generated from autologous skin. *Science, 318:*1920–1923.

Lanza, R. and N. Rosenthal (June, 2004). The stem cell challenge. *Scientific American.* 290(6):92-9. (See also sidebar by Christine Soares.)

Yu, J., Vodyanik, M.A., Smuga-Otto, K., et al. (2007). Induced pluripotent stem cell lines from human somatic cells. *Science, 318:*1917–1920.

BOOKS:

Bailey, R. (2005) *Liberation biology: The scientific and moral case for the biotech revolution.* New York: Prometheus Books.

Fox, C. (2007). *Cell of cells: The global race to capture and control the stem cell.* New York: W.W. Norton and Company.

Herold, E. (2006). *Stem cell wars: Inside stories from the frontlines.* New York: Palgrave Macmillan.

Scott. C. T. (2005). *Stem cell now: From the experiment that shook the world to the new politics of life.* Upper Saddle River, New Jersey: Pi Press.

Questions for Further Discussion or Research

1. National polls often examine stem cell and cloning issues. By searching on the Internet, examine a few questions asked within the past few years to the American public. Poll your class or school, and see if the opinions are similar. Think about making up some of your polling questions. (Be aware that the choice of words used in your question can be very important!) Websites to get you started: http://www.pollingreport.com/science.htm#Stem and http://www.galluppoll.com/.

2. Compare and contrast ESCs and ASCs. Include an explanation of where each cell type comes from and how each type can be isolated. Give examples of how stem cells may be used to help human genetic disease conditions.

3. When first introduced, there was much controversy about *in vitro* fertilization (IVF). Think well into the future when reproductive cloning will presumably be better understood. Consider a time when humans could be successfully cloned and the procedure is safe enough to produce healthy individuals. Is this unethical? Can you compare it ethically to *in vitro* fertilization?

4. In this text, we considered a multipronged approach to using stem cells in treating Parkinson's disease. See if you can find out how scientists are using various types of stem cells for battling some other disease, such as diabetes.

5. If you were about to have a baby, would you consider banking the umbilical cord blood in a private bank, why or why not? See if you can find out the typical costs associated with it. If you decided not to save it for your own personal use, what else could you do with the cord blood?

6. Examine the laws in place for embryonic stem cell research and cloning in the state in which you live. Do you agree with these laws? Why or why not?

7. At the time this booklet was printed, President George W. Bush was still in office. Has a new president had any impact on stem cell and cloning policies? If so, explain.

8. One use for cloning is to make clones of animals, such as pigs, that would be used as a supply for human organ transplants. This area of biology is called xenotransplantation. What are the advantages and disadvantages of xenotransplantation?

9. In the text, we briefly mentioned preimplantation genetic diagnosis (PGD). While this technique can be used to avoid implanting embryos with devastating abnormalities, it can also be used to make "designer babies." Research examples of how PGD has been used and how there may be a fine line between its use in research and in making designer babies. As a side note: *My Sister's Keeper* is an interesting top-selling, fictional novel by Jodi Picoult that examines aspects of the life of a girl who has been chosen by PGD to be a perfect tissue match for her sister with leukemia (based upon a true situation).

10. What is currently being done in stem cell therapies with spinal cord injury?

11. What are some antirejection drugs currently used by tissue and organ transplant patients? What are some side effects of these drugs, and how long can they prevent rejection? Can stem cell therapies avoid these problems with tissue rejection? Explain your answer.